JANE'S POCKET BOOK OF REMOTELY PILOTED VEHICLES:
ROBOT AIRCRAFT TODAY

AF822678

ALSO AVAILABLE IN THIS SERIES:

JANE'S POCKET BOOK of AIRSHIPS
JANE'S POCKET BOOK of COMMERCIAL TRANSPORT AIRCRAFT
JANE'S POCKET BOOK of HOME-BUILT AIRCRAFT
JANE'S POCKET BOOK of LIGHT AIRCRAFT
JANE'S POCKET BOOK of MAJOR COMBAT AIRCRAFT
JANE'S POCKET BOOK of MAJOR WARSHIPS
JANE'S POCKET BOOK of MILITARY TRANSPORT and TRAINING AIRCRAFT
JANE'S POCKET BOOK of MISSILES
JANE'S POCKET BOOK of MODERN TANKS and ARMORED FIGHTING VEHICLES
JANE'S POCKET BOOK of NAVAL ARMAMENT
JANE'S POCKET BOOK of PISTOLS and SUB-MACHINE GUNS
JANE'S POCKET BOOK of RECORD-BREAKING AIRCRAFT
JANE'S POCKET BOOK of RESEARCH and EXPERIMENTAL AIRCRAFT
JANE'S POCKET BOOK of RIFLES and LIGHT MACHINE GUNS
JANE'S POCKET BOOK of SPACE EXPLORATION
JANE'S POCKET BOOK of SUBMARINES

JANE'S POCKET BOOK OF REMOTELY PILOTED VEHICLES: ROBOT AIRCRAFT TODAY

John W. R. Taylor FRHistS, AFRAeS, FSLAET
Kenneth Munson ARAeS

COLLIER BOOKS
A Division of Macmillan Publishing Co., Inc.
New York

Copyright © 1977 by John W. R. Taylor and Kenneth Munson

All rights reserved. No part of this book may be reproduced or transmitted in any form or by any means, electronic or mechanical, including photocopying, recording or by any information storage and retrieval system, without permission in writing from the Publisher.

Macmillan Publishing Co., Inc.
866 Third Avenue, New York, N.Y. 10022
Collier Macmillan Canada, Ltd.

Library of Congress Cataloging in Publication Data
Taylor, John William Ransom.
Jane's pocket book of remotely piloted vehicles.
1. Drone aircraft—Catalogs. I. Munson, Kenneth, joint author. II. Title.
TL671.T34 1977 623.74′69 77-16041
ISBN 0-02-080640-X

First American Edition 1977

Printed in the United States of America

CONTENTS

The first RPV: Professor A. M. Low's original Gnome-engined 'Aerial Target' built at the RFC Experimental Works

FOREWORD

Remotely piloted vehicles (RPVs) represent a whole new concept of aviation, combining the experience gained in more than seventy years of human powered flight with the technological achievements of the missile manufacturers. Like missiles they are not restricted by the frailty of a man in the cockpit; unlike missiles they offer the asset of discretion that can be provided only when a human brain is 'at the controls'.

The first RPVs were target drones, able to manoeuvre far more rapidly than any piloted aeroplane and to give fighter pilots and anti-aircraft gunners live firing practice that tested their skill to the limit. Then, in 1965, the Chinese displayed in Peking what they claimed to be US pilotless reconnaissance aircraft, shot down during clandestine flights over their territory. This was the first glimpse by the public of an advanced type of RPV which could be sent on missions that were too dangerous or politically delicate to be undertaken by men.

From the reconnaissance RPV, carrying cameras and electronic intelligence devices, has evolved the combat RPV. One type carries a laser device with which to mark targets on the ground, so that it can act as pathfinder for squadrons of manned close-support aircraft, or other RPVs carrying 'smart bombs' and homing missiles. Another type has fought an aerial dogfight with a skilled US Navy crew in a Phantom fighter, dodging missiles fired from the Phantom and placing itself in positions from which it could have destroyed the manned aircraft had the combat been a real one.

The Israelis demonstrated the value of RPVs as decoys, to attract enemy anti-aircraft missiles in advance of an attack force of manned fighter-bombers, which could then penetrate the enemy defences while the missile launchers were being reloaded. Less publicised has been the use of 'hero' RPVs to probe the secrets of new, potentially-hostile missile batteries by deliberately getting themselves shot down. Still at the exploratory stage are applications such as the 'seeding' of SIAMs around enemy airfields. SIAMs are fascinating weapons. As long as aircraft based on the airfields stayed on the ground, these Self-Initiated Anti-aircraft Munitions would remain inactive; but they would fire automatically on any aircraft that attempted to take off or land.

These are just a few examples of present and future uses of a new kind of aircraft that may eventually revolutionise aspects of our peacetime life, as well as providing new methods of warfare. And *this* is the first book describing and illustrating the whole family of RPVs not veiled by a 'Top Secret' cloak.

KENNETH MUNSON and JOHN W R TAYLOR

THE FIRST RPVs

SOPWITH

THE FIRST RPVs

Guided missiles and remotely piloted vehicles (RPVs) share a common ancestry, dating back to the first World War. It had taken thousands of years of inventiveness and effort, and had cost countless lives, to devise means of enabling men to fly. Yet, within eleven years of the first officially-recognised piloted flights in a powered aeroplane, by the Wright brothers, European military airmen were already shooting each other out of the sky. Orville Wright shook his head sadly, and commented "What a dream it was; what a nightmare it has become."

Adaptation of the aeroplane for war had been, sadly, predictable. The public would have been more surprised to learn that one man, in England, was soon working to reverse the age-old process by keeping pilots on the ground, had they been allowed to see behind the heavily guarded, locked doors of his laboratory. His name was Professor A.M. Low and he was, in some respects, a 20th century Leonardo da Vinci — conceiving a never-ending succession of brilliant projects in a period before science had advanced sufficiently to make them practicable.

In early 1914, a decade before John Logie Baird's pioneer transmission of a TV picture, Low had demonstrated the principles of television in London. When Britain went to war, he was asked by the Ordnance College at Woolwich to see if the technique could make possible a more effective rangefinder for coastal artillery. Just as success seemed near, Low and his small team of assistants were switched to something entirely different.

Royal Flying Corps (RFC) combat pilots in France had suffered heavy casualties during the German Fokker Monoplanes' period of complete air superiority on the Western Front; their colleagues at home were having little success in their attempts to intercept enemy Zeppelin airships that bombed England by night. Somebody at the War Office had suggested that Low's expertise in radio technology might be used more profitably in developing a remotely controlled pilotless aircraft that could be employed for both interception and ground attack missions.

The basic idea involved packing a small, very simple aeroplane with explosive and guiding it into its target *kamikaze* fashion. This might seem to have made it more missile than RPV. A modern RPV is envisaged basically as an aircraft that is controlled directly but remotely by a human pilot during at least the initial and terminal phases of each flight. It may carry and launch weapons, but is not normally expendable itself. However, these are not inviolable definitions. Often, long-range reconnaissance RPVs have to be pre-programmed for automatic control during deep-penetration sorties over enemy territory. Sometimes, decoy RPVs are employed for operations in a combat area that involve their certain destruction, as a means of saving human lives in a closely following wave of attack aircraft. So it is not too partisan to suggest that Low was the "Father of the Remotely Piloted Vehicle".

Newly commissioned as a Second Lieutenant in the Army, he began his aircraft work in great secrecy in a garage at Chiswick which was allocated the grandiose title of RFC Experimental Works. When progress with the radio control gear justified a move to a more aeronautical environment, the Works was transferred to Brooklands. Sir David Henderson, Director General of Military Aeronautics, had great hopes for what the unit might achieve, and said one day "Now, Low, we must think of some thundering lie to

The Sopwith A.T., which was damaged before completion

conceal the fact that we have a new weapon. Let us call it the A.T.; people will think it is meant as an aerial target."

In the light of later developments, we know that the subterfuge was to have more substance than what it was intended to conceal, as the earliest practical products of the work were to be used as targets. Low could not know this, and a remarkable succession of small experimental aircraft evolved from his experiments.

The first of them, remembered simply as the RFC Experimental Works A.T., had a neat tubular fuselage of original design; but any attempt at overall streamlining was prevented by the need to use components from existing aircraft to save money and time. The wings came from an old biplane; a clumsy skid landing gear was fitted; and the uncowled 50 hp Gnome rotary engine that was mounted on the nose, together with a large-diameter propeller, looked entirely out of scale compared with the airframe. Tests on the ground showed that the engine caused considerable interference with the primitive radio control system. So, although the A.T. was shipped to the Central Flying School at Upavon, it was never flown.

Instead, several of the private aircraft manufacturers of the time were invited to work in parallel with the Royal Aircraft Factory to develop a more practical vehicle for Low's equipment.

One of the first was a small wire-braced monoplane built by Ruston Proctor & Co Ltd of Lincoln. The only surviving information on this A.T. is a single photograph showing it minus engine. By this time designers no longer had to try to utilise anything as cumbersome as a Gnome rotary, as Granville Bradshaw of A.B.C. Motors Ltd had developed from his well-proven 45 hp Gnat the world's first lightweight, expendable aero-engine. It was a small horizontally-opposed twin-cylinder engine, giving 35 hp for a design life of two hours.

One of the six A.Ts built at the Royal Aircraft Factory (now the RAE) to the design of H. P. Folland

Even the famous Sopwith Aviation Company considered it worthwhile taking time out from building fighters by the thousand to design and construct an A.T. around this power plant. The small square-section fuselage box was very like that of the Ruston Proctor A.T., but it had biplane wings and a four-wheel landing gear.

A letter from Professor Low to the author, dated 4 August 1944, recalls that the Sopwith A.T. had a span of about 14 ft (4.27 m). The A.B.C. engine drove an ordinary wooden propeller, about 3 ft 6 in (1.07 m) diameter, and the nose also housed an explosive charge weighing about 50 lb (22.5 kg). Behind that were fuel and oil tanks, ignition batteries and other equipment. Then came a box with a glass lid, slung on rubber supports. Measuring about 2 ft 3 in by 9 in (0.69 x 0.23 m), this contained all the relays, receiver and the Key system which prevented interference. Aerials were intended to go inside the wings, but can be seen wound around the rear-fuselage and wingtips on the prototype. In the rear-fuselage, driven by a shaft from the engine, was a mechanical relay, so that each contact made in the control box caused the engine power to operate the control surfaces (which may have been limited to rudder and elevators). In other words, the Sopwith A.T. was designed for full servo control in 1916.

It was damaged before completion, and abandoned, although Sopwith completed and flew a somewhat larger aircraft, known as the Sparrow, of similar configuration but with a normal cockpit for a pilot.

Geoffrey de Havilland seems to have persisted longer with the

little monoplane A.T. which he designed around the A.B.C. expendable engine. No details survive, although this may have been the A.T. which Low attempted to fly at Upavon on 21 March 1917. The late C.G. Grey, who succeeded Fred T. Jane as Editor of *All the World's Aircraft*, told of how representatives of all the Allied and friendly powers were invited to Upavon to watch a demonstration of Britain's new secret weapon on this or a later occasion. Attachés, dressed in gorgeous white uniforms with medals and gold braid, scuttled ignominiously in all directions when the A.T. misbehaved after launch and dived towards them. Major Gordon Bell, a pilot remembered as one of the "characters" of an age of larger-than-life officers, turned away calmly with the comment "I could throw my b..... umbrella further than that".

Nor did the Royal Aircraft Factory itself achieve any great success. One of its leading designers, H.P. Folland, who was responsible for the S.E.5 fighter, evolved simultaneously a simple monoplane A.T., spanning 22 ft (6.70 m) and powered by the 35 hp A.B.C. engine. There were no control surfaces on the wings, which were rigged with considerable dihedral to ensure stability without any need for gyros. Aerial wires were wound around the rear-fuselage and chordwise around the wings.

At least six of these Farnborough A.Ts were built during 1917. As Low's Experimental Works had moved again by then, to Feltham, the flight testing was conducted at Northolt aerodrome, a few miles to the north. A launch track was laid down, consisting of an initial 50 ft (15 m) horizontal run followed by a 100 ft (30 m) incline.

Artist's impression of the take-off of America's 1917 'flying bomb' developed for the US Navy by the Curtiss, Sperry and Delco companies

The first flight, on 6 July 1917, was brief. Under radio control, the A.T. became airborne about halfway along the track, climbed steeply until it stalled, and then crashed. The next tests, on 25 and 28 July, were made with the elevator fixed at a predetermined setting to prevent any possibility of over-control from the ground. This invited trouble, as it left the operator with minimal control of the aircraft once it had left the track. Both tests ended disastrously, and there was little further A.T. activity before the end of the "war to end wars" made such weapons redundant.

Over in the United States, there had been equally brief and inconclusive efforts to produce what were called, more openly, aerial torpedos. First of them was a 600 lb (272 kg) pilotless biplane built under Navy contract by Curtiss, to carry Sperry and Delco equipment. The prototype was delivered to the Sperry Flying Field at Copiague, Long Island, on 10 November 1917. Powered by a 40 hp Ford engine, it was intended to take off from a four-wheel carriage running on two rails, and was not remotely controlled from the ground. The launch crew was expected to estimate the distance to the target and, after allowing for wind drift, point the "flying bomb" in the right direction and set the engine to run for the appropriate time to bring the weapon over its objective.

In flight, the aircraft's direction of flight was controlled by a simple gyroscope and its altitude by an aneroid barometer. Over the target, the engine was supposed to stop, after which a mechanical device extracted bolts holding the wings in place. The fuselage then dropped on to the target — hopefully — complete with 300 lb (136 kg) of explosive.

The US Army Air Service had a counterpart of the Navy's aerial torpedo, known as the Dayton-Wright Kettering Bug. This was a neat biplane, powered by a four-cylinder Vee engine, and was

RAE 1921 target on its launch ramp on board HMS Stronghold (Crown copyright)

Larynx – the 'Long-range Gun with Lynx engine' – awaiting launch from HMS Stronghold

designed to carry a 180 lb (82 kg) explosive charge a distance of 40 miles (64 km) at 55 mph (88 km/h). Tests were terminated at the end of the first World War, but a replica is among the most popular exhibits in the US Air Force Museum at Dayton, Ohio.

Being entirely self-contained after take-off, these two US aircraft were missiles rather than RPVs. To pick up again the main thread of RPV development we must return to the UK.

Economic problems left by the war, and the expectation of a long period of peace, meant that the Royal Air Force was not allowed to spend money on buying new aeroplanes to replace its wartime types. This may explain why interest in low-cost, pilotless alternatives was revived as early as 1920. Three categories of aircraft were proposed:

1. a gyroscopically-controlled "ammunition carrier" intended to fly on a steady course at a constant speed for a pre-determined distance.
2. an inherently stable "aerial target" with a range of 20 miles (32 km).
3. an "aerial torpedo" which could be powered or unpowered, and would be radio controlled for 10 miles (16 km) after launch from an aircraft.

Confidence that such machines could be perfected might seem misplaced after years of wartime effort had produced not one operational A.T. In fact, much had been achieved. Low had devised a technique by which the "flying bomb" could be controlled over long ranges by a radio operator in an accompanying parent aircraft. The final version built at Farnborough had not needed a specially laid take-off track, being launched by a compressed air catapult mounted on a truck. No live warhead was ever fitted; but the point of impact was signalled by automatic ignition of a petrol flare. In themselves, each of these developments may seem trivial; together they pointed the way to future, operational RPVs.

The first of the new generation was not very different from the 1917 A.T. It was decided to abandon the third of the proposed designs, the "aerial torpedo", and to combine the other two in what was designated the RAE 1921 Target, the initial letters standing for Royal Aircraft Establishment, where the development was to be centred. Airframe configuration differed from the 1917 A.T. mainly in deletion of the upper tail fin and landing gear. Wing span was 23 ft (7.01 m), length 18 ft (5.49 m) and launch weight 630 lb (285 kg). Powered by a 45 hp Siddeley-Deasy Ounce, the "Target" was expected to carry a 200 lb (90 kg) warhead at a speed of 103 mph (165 km/h) at 6,000 ft (1,830 m).

Early hopes were dashed on 31 August 1922 when the first test aircraft, launched from a simple track on the deck of the aircraft carrier *Argus*, fell straight into the sea. The second test, a few days later, was no more successful. To increase initial flying speed, a launch ramp was mounted on a destroyer, HMS *Stronghold*, up which the Target could be accelerated by a large bag full of water, which dropped vertically into the sea. Four trial launches in 1923-24 showed that more work was necessary before the gyroscopic control mechanism could ensure a stable climb-away. The decision was taken to fit full radio control in place of the clockwork mechanism that had been employed to actuate the controls in pre-programmed sequence.

From that moment, there was a dramatic improvement. The seventh launch, from the *Stronghold*, on 3 September 1924 demonstrated that stability and control were now entirely

Fairey Queen, the first of a long series of full-scale pilotless target aircraft extending to the present day

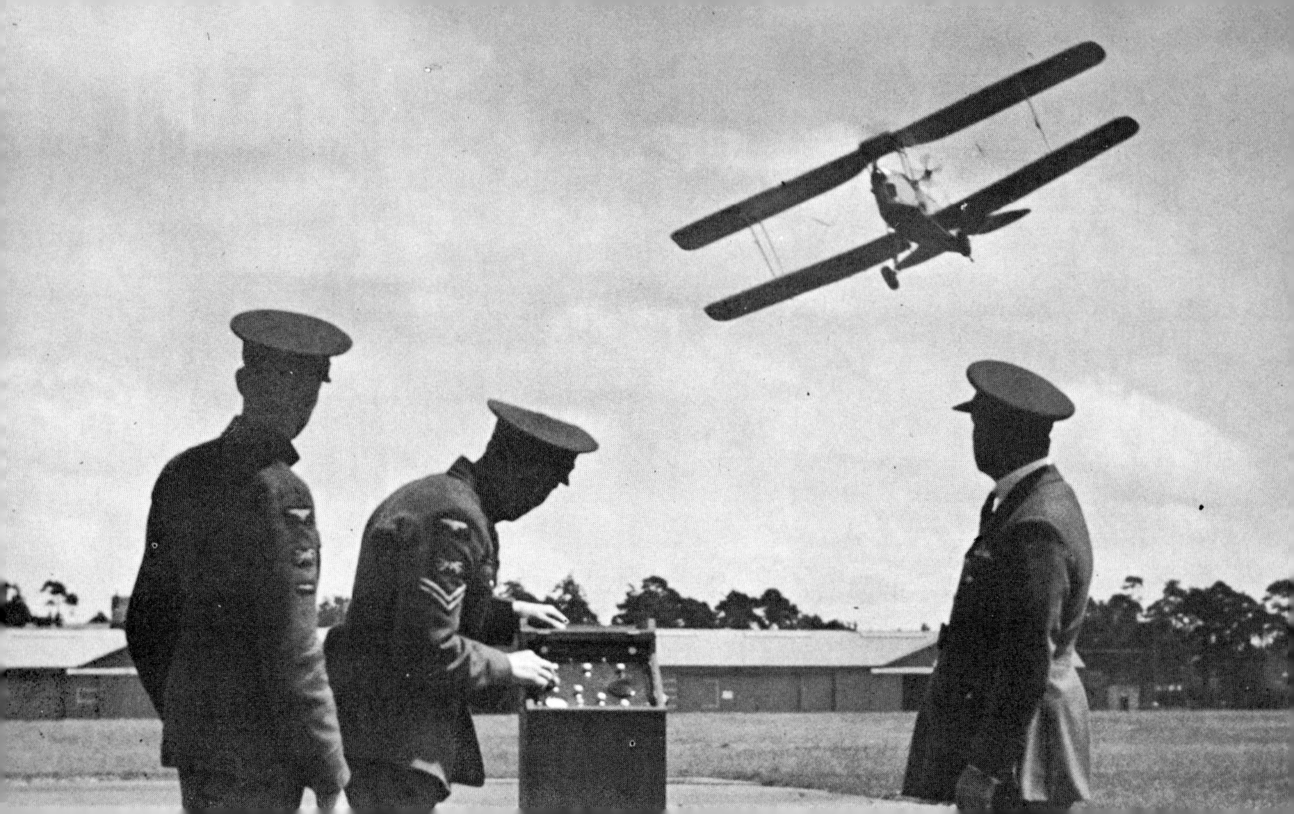

satisfactory; the flight terminated after twelve minutes only because the engine stopped. Duration of the tenth flight, in the following February, was 39 minutes, during which the aircraft responded to 43 separate command signals. This was regarded as sufficient justification of the concept to advance to a more formidable production-type weapon that would carry a 250 lb (113 kg) charge for a distance of 300 miles (480 km). The design was refined, the end product being a clean little mid-wing monoplane with a cruciform tail unit, no landing gear, and a 200 hp Armstrong Siddeley Lynx engine close-cowled in a low-drag cowling at the front of a tubular fuselage. Its designed maximum speed of 193 mph (310 km/h) was much higher than that of contemporary fighter aircraft; so the authorities were particularly unimaginative when they named it "Long-range Gun with Lynx Engine" — soon abbreviated to Larynx.

Last aircraft designed and constructed entirely at the RAE, it was years ahead of anything comparable conceived in other countries. Of twelve built at Farnborough, the first suffered a mechanical failure shortly after leaving its cordite-operated catapult on HMS *Stronghold*, on 20 July 1927. The next attempt was more satisfactory, the third an outstanding success. Larynx flew the full length of a course over the sea, off the coast of Somerset, Devon and Cornwall. Numbers four and five were also launched from the *Stronghold*, number six from a catapult ashore, on Portland Bill.

Next stage was to test Larynx with a live warhead; so five of the six remaining aircraft were taken to the RAF Station at Shaibah, in Iraq, where their launch catapult was installed so that they could deliver their warheads harmlessly in the desert. One of the ensuing launches, made between 31 August and 3 October 1929, was a failure. Engine and fuel system troubles brought three others to a premature end. Only the fourth Larynx disappeared into the distance, never to be seen again.

First full-scale radio-controlled target aircraft produced in large numbers was the de Havilland Queen Bee, based on the Tiger Moth trainer. It is shown here about to fly over its control console

Had it been a true RPV, with radio control, its fate might have been known; but Larynx employed an automatic pilot — as did the German V.1 flying bombs which eventually proved the practicability of the concept in 1944-45 and pointed the way to the jet-powered, nuclear-warhead cruise missiles of today.

Whether or not anyone had a guilty conscience about where the fourth Iraqi Larynx might have ended its flight, and deposited its lethal load, we may never know. From that time, however, RPV research in Britain was directed towards the development of radio-controlled targets — intended, honestly and solely, to be shot at, and not to perform any lethal function themselves in the intended manner of the A.Ts of the first World War.

The requirement was sparked off by a dispute between the Royal Air Force and Royal Navy concerning the effectiveness of bombing and torpedo attacks by aircraft against battleships. Recalling the sinking of several old warships by Martin MB-2 twin-engined heavy bombers of the US Army Air Service, in tests organised by General "Billy" Mitchell in 1921-23, RAF leaders insisted that capital ships were too vulnerable to survive in war. The Navy, with some justification, pointed out that the bombing might have achieved less had the ships been manoeuvring and able to shoot in self-defence.

To resolve the argument, in October 1930 the Air Ministry requested the RAE to produce a radio-controlled target aircraft which could be used for simulated attacks on the Fleet in the teeth of every gun that the ships' crews could bring to bear on it. Three standard Fairey IIIF reconnaissance floatplanes were modified as "Fairey Queens". Changes were minimal, being limited to installation of the radio control gear and an increase in wing dihedral to ten degrees, to enhance stability. Check flights with a pilot on board were made in the Autumn of 1931. Without pilots, the first two aircraft crashed almost as soon as they left the catapult launcher on HMS *Valiant*, on 30 January and 19 April 1932.

If the Navy felt any malicious glee at the failure of the would-be "attackers" this was to be short-lived. After a successful nine-minute flight under radio control on 14 September, the remaining Fairey Queen was launched for use as a target by the Home Fleet during its Spring cruise in the Mediterranean, in January 1933. It survived more than two hours' concentrated gunfire before being recovered safely; although it was shot down by gunners on HMS *Shropshire* four months later, the Admiralty had learned a valuable lesson.

Specification 18/33 was drawn up rapidly to cover development of a radio-controlled all-wood target version of the de Havilland Tiger Moth biplane, as a relatively inexpensive means of giving Fleet gunners the practice they so clearly needed. The result was the famous Queen Bee, of which 420 were built for use by the Navy and Army, between 1934 and 1943. The best testimony to their value is that enormous numbers of other full-scale radio-controlled target aircraft have been built since they spent their lives in skies full of shrapnel and shellburst, including modern jets (illustrated in this *Pocket Book*) able to fly at two and a half times the speed of sound.

SMALL, REALISTIC, SURVIVABLE . . . . .

SMALL, REALISTIC, SURVIVABLE.....

Most observers of pre-war Fleet exercises have memories of Queen Bee target aircraft that seemed to fly for hours ahead of black puffballs of smoke from exploding anti-aircraft shells. They proved surprisingly difficult to destroy. Nonetheless, occasional newspaper photographs of bedraggled shot-down aeroplanes being hauled out of the water by crane could only cause heartsearching in an age that still echoed with words like depression, unemployed and poverty.

The development which cut costs enormously had its origin, surprisingly, in Hollywood — the synonym of fairytale luxury and vulgar extravagance, rather than economy, in the 'thirties.

Film star Reginald Denny was well known as a model aircraft enthusiast of the most professional kind. It seemed to him a waste of good money to shoot down full-scale Tiger Moths. So, in 1935, he built a 9 ft (2.75 m) span model, of conventional braced high-wing monoplane configuration, with a normal fixed tailwheel landing gear and powered by a small piston engine. Maximum speed was only 50 mph (80 km/h), compared with the Queen Bee's 109 mph (175 km/h); but the lower cruising altitude of the model offset the speed differential so far as the gunner was concerned.

Denny designated his little prototype the RP-1 and formed the Radioplane Company to handle development and future production, if he succeeded in interesting the US services in his idea. The Army agreed to let him demonstrate the RP-1 at Fort McArthur, but showed little interest after it crashed. Radioplane persisted in its experiments, and by 1939 was flying the RP-4. The basic airframe looked little different from that of the original prototype, but the single engine now drove two contra-rotating propellers mounted, uniquely, side by side. Take-off was from a twin launch-rail, with a wide-track, low-pressure tricycle landing gear to cushion landings on the Californian desert.

The Radioplane RP-1 target aircraft of 1935, built by film star Reginald Denny (right). This was the ancestor of all modern American target drones

The RP-4, which had a steel-tube fuselage, soon proved that it was no toy. During tests it flew at 60 mph (96 km/h) and could be controlled over a range of 800 ft (245 m). The US Army Air Corps ordered a few. Then, in 1942, the United States found itself at war; suddenly, Radioplane received contracts from the Army and Navy for all the targets it could turn out, as did the Frankfort company. By the end of the war, they had delivered 984 OQ-2As, 9,403 OQ-3s and 3,548 OQ-14s. All three types spanned about 12 ft (3.66 m). The -3 and -14 each had a 20 hp Righter engine and could reach 140 mph (225 km/h) and 10,000 ft (3,050 m).

Today Radioplane is the Ventura Division of the Northrop Corporation, and has delivered over the years more than 76,000 pilotless targets to the US military services and 25 allied nations. Most have been piston-engined, but Northrop Ventura has also found a ready market for its MQM-74 jets, not least in Israel.

For obvious reasons, the Israelis are reluctant to disclose details of the combat applications they devised for what had been supplied as jet targets, before and during the Yom Kippur War of 1973. However, it is known that Northrop Chukars (export versions of the MQM-74) were despatched in advance of some waves of piloted attack aircraft. They saved many lives by appearing as much larger aircraft on enemy radar screens, causing the Arab

First entirely successful US pilotless aircraft, the RP-4, built by the Radioplane Company (now the Ventura Division of Northrop Corporation) in 1939

Radioplane's OQ-3, built in large numbers during the second World War, was the first to embody a target recovery system

forces to fire their missiles at the Chukars and so allowing the main attack formations to slip through the defences while the missile launchers were being reloaded. A high proportion of the RPVs were shot down — múch to the dismay of the Israelis, who developed such a high regard for them that the loss of every one hurt.

In case any future enemy might devise a method of distinguishing decoy RPVs from larger, piloted aircraft, companies in Britain and the USA are working on what are called harassment RPVs. These perform the same basic function as the decoys but carry anything from a 2 lb (1 kg) explosive charge to a miniature atomic warhead, aimed at specific battle area targets in such a way that they cannot be ignored.

There are many interesting variations of this category of pilotless vehicle, some still highly secret. Occasionally, one hears of "hero missions" performed by RPVs, implying mechanical rather than human suicide. All that may be published is that such missions have sometimes taken the vehicles over important missile sites. Any that was allowed to pass unhindered might have carried cameras and other sensors able to record details of the sites. Any RPV brought down by one of the missiles might well have contained electronic devices able to telemeter to its base, in fractions of a second, invaluable data on the guidance techniques, speed, size and other characteristics of the weapon that was about to destroy it.

The RPVs used to date on "hero missions" have also been evolved from targets, by the company whose founder, Claude Ryan, built Lindbergh's *Spirit of St Louis* transatlantic aeroplane in 1927. Teledyne Ryan began flight testing its original Firebee I jet target in 1951, and has since manufactured nearly 6,000 production models. Current versions have a maximum speed of Mach 0.96, can make precision low-altitude flights only 50 ft (15 m) over water, and remain airborne for well over an hour, climbing above 60,000 ft (18,300 m) if required. They can be controlled from another aircraft or a ground station, can pull 6*g* in simulated air combat, can be fitted with electronic countermeasures (ECM) devices and signal jammers to make detection difficult, augmentors to make them look as large as a piloted combat aircraft, and a whole range of other equipment. They can even tow a small target at the end of a cable, so that a missile or anti-aircraft shell that is on target will not destroy the Firebee itself but only the towed "bird".

Evolution of this versatile target drone into an operational RPV is reckoned to date from the time of the Cuban missile crisis in 1962. An American Lockheed U-2 spy-plane and its pilot were shot down during a vital reconnaissance flight over Cuba. Its mission over the mushrooming missile sites could have been performed equally efficiently by a remotely controlled unmanned aircraft. So, with only two more U-2s immediately available to keep an eye on such trouble-spots, the US government initiated an urgent programme to develop RPVs capable of supplementing them.

Within 90 days Ryan had produced its first Model 147 RPV, based on the Firebee I. Under a programme code-named Big Safari, production Model 147As were tested first against America's own sophisticated air defences. They were able to make repeated

Ryan Q-2A Firebee targets under the wings of a Douglas A-26 from which they were air-launched as targets for missiles and rockets fired by USAF fighter pilots during exercise William Tell II. From this original Firebee have evolved many of the Ryan RPVs and targets described and illustrated in this Pocket Book

penetrations without detection, and became the first members of a family of front-line RPVs that now numbers more than 20 identifiable variants.

It is indicative of the sensitive nature of their work that the public knows so little about them. If so-called spy-planes discover missiles being emplaced in somewhere like Cuba...if a pilot named Gary Powers gets himself shot down over territory where he has no right to be...if an electronic intelligence (elint) gathering aircraft is attacked by enemy fighters near "the other side's" border or coastline, the incident makes big news for a few days, and is then forgotten. But such flights are not isolated incidents. Peacetime reconnaissance and electronic intelligence gathering are a facet of everyday life, routine but clandestine.

If an intruder aircraft, shot down over a foreign nation, contains no human crew, even the formalities of a diplomatic protest are seldom observed. Thus, the first indication that RPVs had superseded U-2s for certain missions came when the Chinese displayed in Peking the remains of unmanned reconnaissance aircraft that had been intercepted during illegal overflights of their country. It was not difficult to identify the row of wrecked airframes as long-span, high-flying variants of the Firebee I target; and gradually the pieces of the jigsaw began to fit together. In particular, the purpose of Lockheed DC-130 Hercules drone-launching director aircraft seen at operational airfields in countries bordering places like China could be deduced.

Since the mid-sixties the United States has invested several hundred million dollars in RPV programmes, although most of its military leaders have yet to recognise the full potential of pilotless aircraft in a combat environment. They acknowledge that Ryan 147s did superb work in Vietnam, photographing targets for air attack, recording damage after bombing, even discovering unsuspected key targets like the huge North Vietnamese fuel storage area in a suburb of Hanoi. Equally, it must be admitted that some of the most valuable "bonus" discoveries resulted from the preprogrammed aircraft wandering off course.

The 147s flew over North Vietnam at both high and low level, relying on their speed, small size and ECM equipment to elude the tremendously heavy and efficient defences. Typical of the equipment developed for such aircraft is the Itek KA-80A optical bar panoramic camera, with a focal length of 24 in (610 mm). As the RPV flies over a target area, this 325 lb (147 kg) camera can take 1,500 superbly detailed photographs, each 45 in (1,140 mm) long by 4.5 in (114 mm) wide.

Specially funded programmes during the period of the Vietnam war extended the usefulness of the Ryan 147s far beyond "hit and miss" photographic reconnaissance. The Type 147TE was produced for the USAF's Compass Dawn and Compass Dwell programmes, aimed at the precise pinpointing of enemy ground radars. The 147NA, built under Compass Bin and Combat Angel programmes, was a medium-altitude ECM version, carrying active (jamming) or passive (chaff dispensing) pods. Such RPVs could replace manned ECM escort aircraft protecting an air strike force.

The Ryan 234/259 family, also developed from the basic Firebee I, is far from protective, being intended primarily for tactical strike and defence suppression roles. The US Navy has even evaluated RPV interceptors in simulated air combat against a McDonnell

Early tests to evaluate the attack capability of RPVs were made with a modified BQM-34A Firebee, here seen releasing two inert 500 lb bombs over White Sands Missile Range, New Mexico

Douglas F-4 Phantom II flown by a highly experienced aircrew. The results were thought-provoking. All the Sparrow and Sidewinder missiles launched from the Phantom missed the RPV, but the latter manoeuvred itself several times into positions from which it could have brought down the Phantom had it been armed.

For the present, little work is being done on interceptor RPVs. Top priority is being given to perfecting a small, low-cost "mini-RPV" fitted with a laser designator and a TV camera by which the pilot on the ground can guide it towards a target, such as a tank or strongpoint in a combat area. Having located the target on his TV screen, the pilot can "lock" on to it both the TV "eye" and the laser designator. The RPV is then set to circle the target area, and the strike force of piloted aircraft is directed towards it. Laser-guided missiles under the wing of each aircraft are fitted with seeker heads which pick up and lock on to laser signals being bounced off the target by the RPV's designator. This enables the missiles to home unerringly on the target to destroy it.

This is only a first stage towards what RPV experts see as the automated battlefield of the future. Instead of the strike force of piloted aircraft, there is no reason why the target should not be hit by laser-guided artillery, laser-homing ballistic missiles, or even air-to-surface laser-guided missiles carried by a formation of RPVs. Move the scene out to sea, and the target-designating mini-RPV could be launched underwater from the torpedo tubes of a submarine. It could then find and mark surface ships, for attack by cruise missiles or ballistic rockets launched from the submerged submarine. Such a combination would threaten the entire future viability of surface fleets — provided no sure way is found of locating submerged submarines by means of long-endurance ocean surveillance RPVs, or satellites.

Pioneer 'mini-RPV' was Teledyne Ryan's Model 262. Made of glassfibre, and with hardly a straight line or flat surface to reflect radar signals, it would appear to be no larger than a very small bird on the screen of an enemy radar operator

ROBOT EYES

ROBOT EYES

The advantage in war has always been held by the commander who had a means of knowing what was happening "over the hill" on land, or beyond the horizon at sea. Spying, intelligence gathering, reconnaissance by cavalry and aircraft, observation by balloons and satellites, these and other techniques have been used at some time or other during the thousands of years since a group of men first decided to acquire by force the possessions of their neighbours. To older-established reconnaissance devices, the war in Vietnam added RPVs.

Photographs on this and the following pages show how the Ryan 147s performed their unpublicised but vital tasks, often flying in daylight through some of the most heavily defended airspace in the world to bring back their photographs. The sequence illustrating preparation for a sortie, take-off, launch and recovery shows a number of different RPVs, on different occasions. Even now it is not permissible to trace the story of one, typical, genuine operation.

Use of RPVs in Vietnam provided US commanders with constantly-updated information which could have been acquired only at great risk to American lives had not the little aircraft been available. This was particularly true of photographs taken during the final massive USAF "Linebacker II" bomber offensive, between 18 and 30 December 1972, which created the conditions for US disengagement from a costly war that had gone on for too long.

Ryan 147NC (AQM-34H) medium-altitude ECM RPVs undergoing pre-flight checkout

Fitting ALE-2 underwing chaff-dispensing pods to an AQM-34H. By dropping over enemy territory slivers of metal foil (like the 'Window' dropped by the RAF in the second World War), such RPVs confuse defensive radars and so protect the strike forces for which they provide escort support

With a reconnaissance RPV already in place under its wing, a Lockheed DC-130 Hercules launch/director aircraft waits for take-off time on an airfield in Vietnam in the early 'seventies. Each Hercules could carry up to four Ryan 147 RPVs

Above: *As darkness begins to fall a DC-130 heads for its designated launch area with a high-altitude night reconnaissance RPV clutched under its starboard wing*

Right: *Without crossing the border of enemy territory, and beyond the range of missile defences, the DC-130 releases one of its RPVs – in this case a Model 147TF (AQM-34R) surveillance drone*

Left: *Diagram showing the way in which a USAF recovery helicopter retrieves a reconnaissance RPV at the end of its sortie. After the RPV has deployed its on-board parachute, the helicopter lowers a trapeze structure with which it snares a small drogue chute above the main canopy.* Above: *Back at base, in Vietnam, the HH-3E lowers its charge gently to the ground, where technicians wait to process within minutes the film in its camera pack*

PILOTLESS PR

On this and the two following pages are examples of the photographs taken as routine by Teledyne Ryan photographic reconnaissance (PR) RPVs during the war in Vietnam. Both medium and low altitude sorties are represented, with film from vertical and obliquely mounted cameras. The quality is much lower than that of the prints available to US intelligence personnel, as our printing plates are not only 'screened' but were made from copy prints supplied by the Department of Defense. Of particular interest are the 'Before' and 'After' pictures on page 44, as they depict the way in which the RPVs could be directed back precisely to a previously-photographed target area after a bombing raid, to report the extent of the damage. The accuracy of the attack, in this instance, is indicated by the limited damage to houses and other buildings adjoining the primary target.

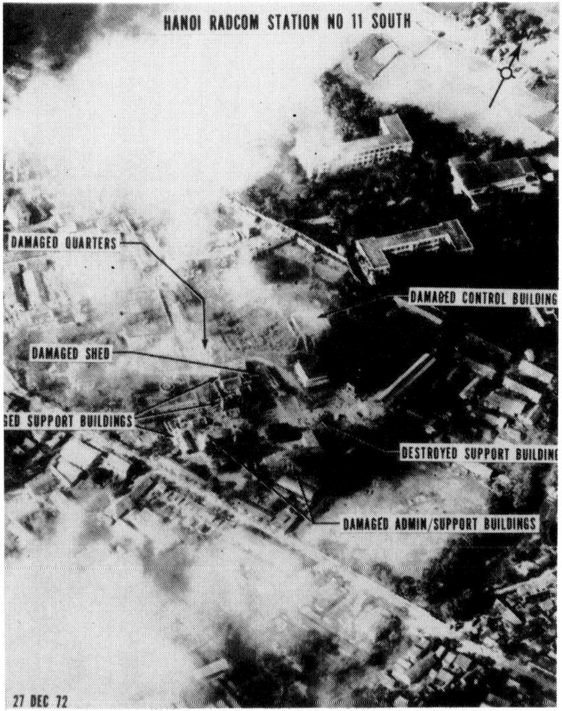

Medium-altitude photograph, taken through broken cloud, showing the damaged state of a radio communications centre in Hanoi, North Vietnam, on 27 December 1972

HANOI THERMAL POWER PLANT
BEFORE

HANOI THERMAL POWER PLANT
AFTER

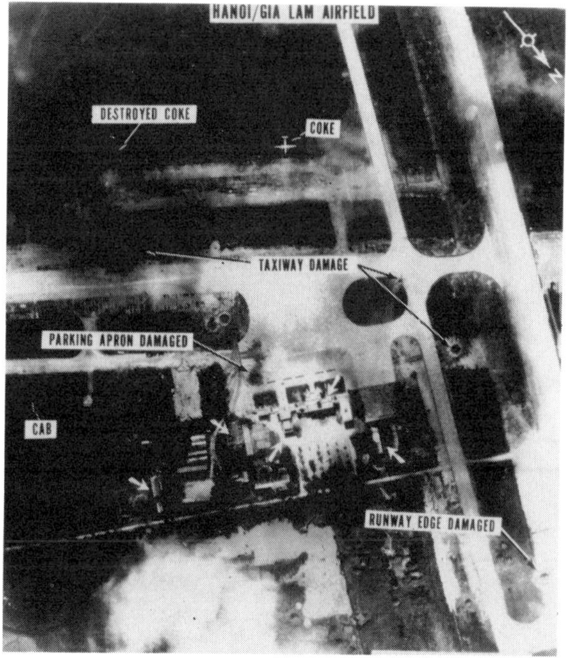

Right: *Aircraft identified by the western reporting names* Cab *and* Coke *on this photograph were, respectively, Lisunov Li-2 (licence-built DC-3) and Antonov An-24 transports*

MISSILE-AGE A.T.s

U.S. AIR FORCE
D4
MJ11

MISSILE-AGE A.Ts

The very first RPV built by Professor A.M. Low was, as we have seen, intended as a flying bomb, although it was designated A.T. to suggest it was merely an aerial target. The wheel has now turned full circle. From modern Firebee aerial targets has been evolved a formidable family of ground attack RPVs designated Ryan Models 234 and 259 (see pages 210-13). This series of photographs shows a flight test of a Model 234. Unlike guided missiles, which are expendable, the RPVs make their attack with other weapons and are themselves recovered at the end of each flight, for repeated use.

Left: *A HOBOS TV-guided missile is loaded on to the Ryan 234 (BGM-34A) RPV, which is already in place on the underwing pylon of a DC-130 Hercules*

Above: *HOBOS exists in many forms. This one consists of a TV homing head and aft control section attached to a standard bomb*

Left: *As the time for launch approaches, the drone controller on board the DC-130 prepares to guide the RPV towards the target area*

Above: *Under the power of its own turbojet engine, the Model 234 begins a shallow dive at very high subsonic speed, making it a difficult target for the ground defences. This one is carrying a Maverick missile*

Above: *Already the operator in the launch aircraft has located the target by means of the TV eye in the nose of Maverick, and has 'locked' the missile on target. The RPV can now be recalled for retrieval.* Right: *This remarkable ground-level photograph shows the Maverick nearing its objective, an old radar van.* Far right: *Lacking a live warhead, the Maverick passes clean through the centre of the radar van and emerges from the far side*

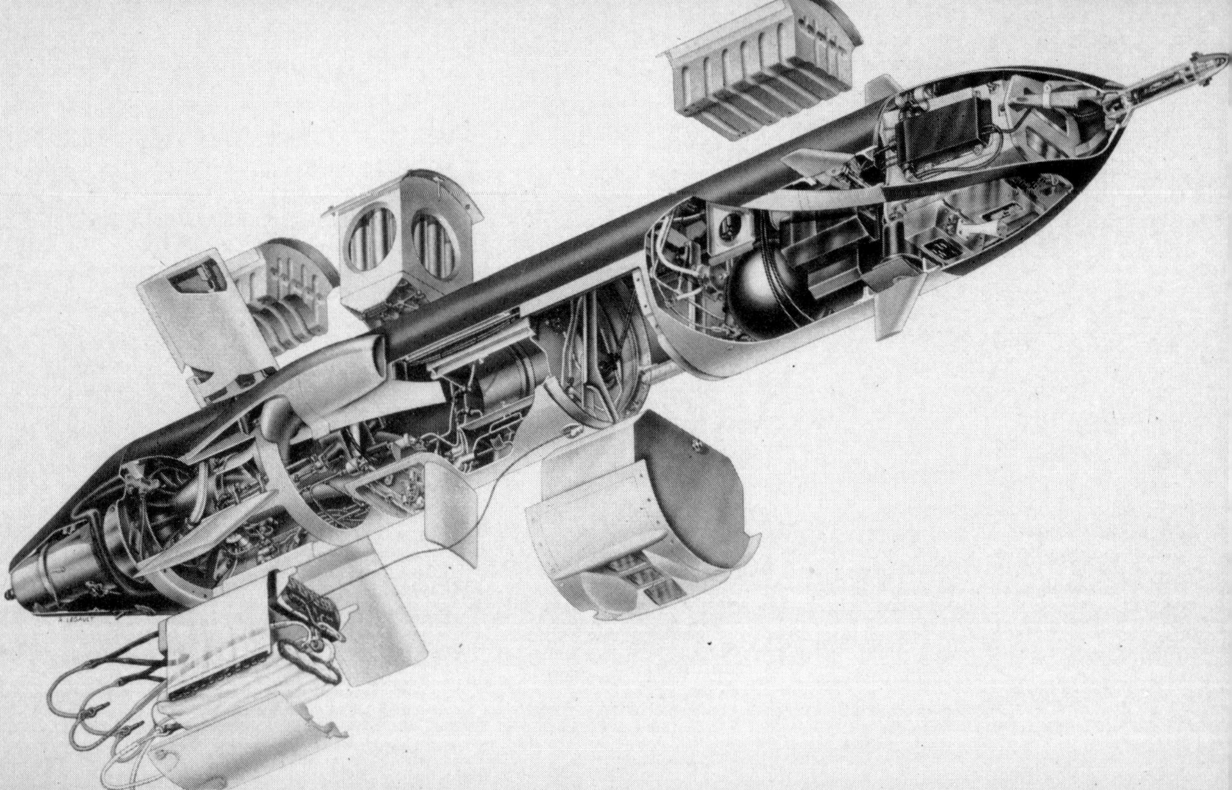

Far left: *Exploded cutaway drawing of the Canadair AN/USD-501 surveillance RPV, showing (at top) the forward and rear landing bag containers and flare pack, and (underneath) the centre-fuselage sensor pack and parachute recovery pack.* Left: *A large map indicates the course to be followed by an AN/USD-501 during a demonstration flight from Camp Shilo, Manitoba, Canada.* Above: *Folding back the side and roof covers that protect the RPV and its launcher during transport*

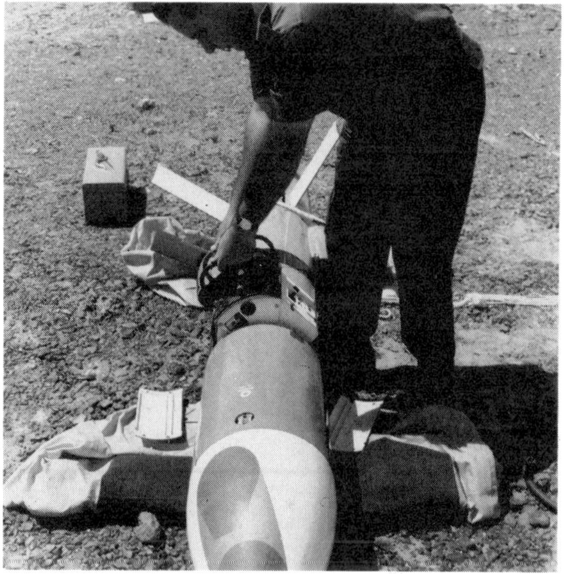

Far left: *The RPV leaves the short ramp on the transport/launch vehicle, with the aid of a rocket booster which jettisons after burn-out.* Left: *The RPV is inverted for recovery at the end of a flight, so that the inflated landing bags can cushion the touchdown.* Above: *Removing the sensor pack. Films are rushed to a mobile facility equipped with newly developed rapid processing systems*

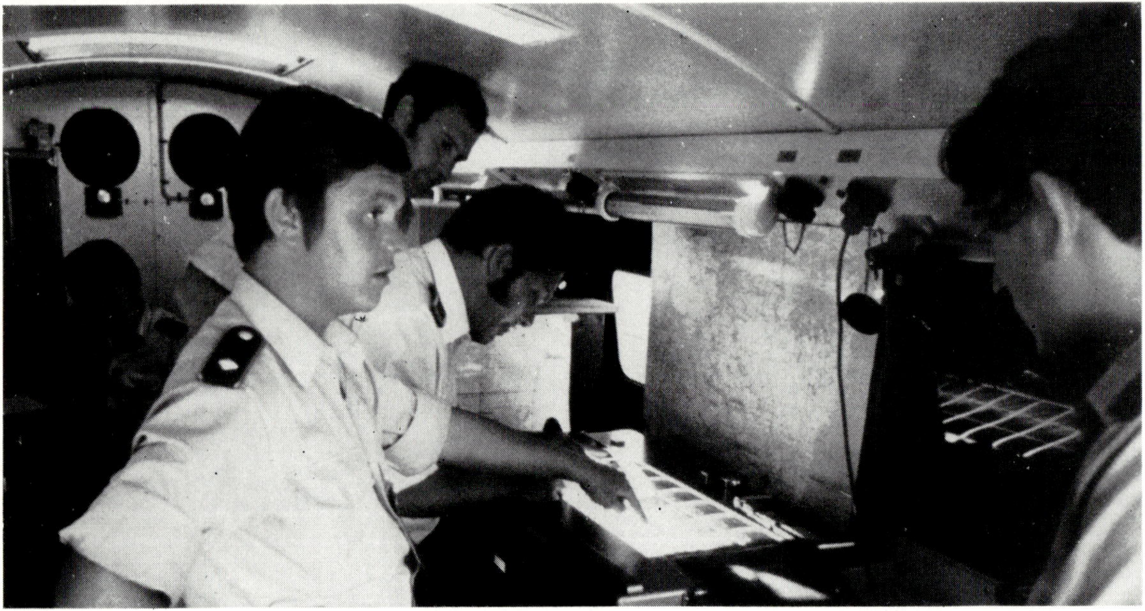

Interpreters discussing details shown on photographs brought back from a typical sortie by an AN/USD-501

Above: *USS* Blandy, *the destroyer equipped for sea trials of the Eglen Falcon multi-role RPV under the Naval Surface Weapons Center's Ship Tactical Airborne RPV (STAR) programme.* Right: *Two men have no difficulty in handling the mini-RPV, even on the cluttered aft deck of the ship*

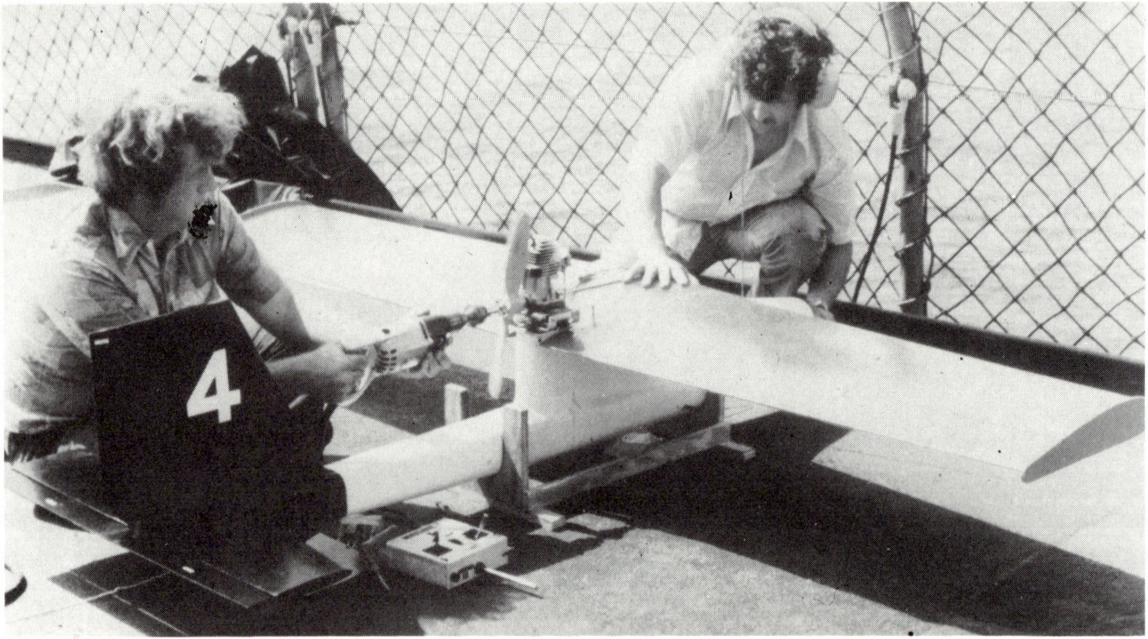
4

Left: *Starting the engine of the Falcon*

Above: *Pneumatic launcher installed on the* Blandy *for the STAR trials*

Left: *Falcon approaches the capture net rigged on the aft deck of the destroyer.* Above: *Removing the RPV from the recovery net at the end of its sortie*

RPV DATA SECTION

AEL4111 SNIPE

Target system
Data apply to target towing aircraft
Power plant: One 56 cc piston engine
Wing span: 8 ft 3 in (2.51 m)
Length overall: 7 ft 0 in (2.13 m)
Payload: 10 lb (4.5 kg)
Max launching weight: 40 lb (18 kg)
Max speed: 111 knots (128 mph; 206 km/h)
Max controllable range: 2.6 nm (3 miles; 4.8 km)
Average flight endurance: 40 min (extendable if additional fuel tanks fitted)
Launch and recovery: Launched by catapult from trailer-mounted ramp. Recovery by parachute or by orthodox landing. Fail-safe system shuts down engine and deploys parachute in event of radio interference or loss of control signal.
Guidance and control: Radio command guidance system. Each Snipe customer must specify two control frequencies in the 27MHz or 70MHz waveband which are clear of all other traffic. Each system operates in a frequency slot with an approx bandwidth of 25KHz and with an ERP maximum of 5W. If it is intended to fly more than one aircraft simultaneously, at the same site or within radio frequency control range of each other, individual control frequencies must be specified for each set. If such coincidental operation is not intended, the customer's total order can be supplied to operate on the same control and reserve frequencies.

Complete Snipe system, including trailer, launcher, two aircraft, towed target and optical tracking gear

Special equipment: The complete Snipe system comprises two miniature aircraft, including their engines, airborne guidance equipment, visual enhancement high-intensity strobe lights and radio-triggered recovery parachutes; tripod-mounted ground control equipment, complete with optics; a ground support unit, including engine starter, fuel container, accumulator and battery chargers; radio monitor; a field spares pack, including a set of key replacement subassemblies and components; a complete target drogue assembly, including a 6 ft (1.83 m) x 2 ft 0 in (0.61 m) banner and 1,000 ft (305 m) towline; five spare drogues; and a lockable trailer for containing, storing and transporting the complete system. The system catapult launcher forms an integral part of the trailer. Optional extras include additional fuel tanks or an additional 10 lb (4.5 kg) of payload; and underwing smoke pods for use with Simflak AA laser system.
Remarks: AEL also produced earlier the **MATS-A** (Model Aircraft Target System A) currently in use by the British Army. This has now been superseded by the **AEL 4560 Smidge**, an aerial small arms target of similar concept with a 7cc engine, wing span of 5 ft 6 in (1.68 m), length of 4 ft 5 in (1.35 m), gross weight of 5 lb (2.3 kg), speed of 69.5 knots (80 mph; 129 km/h) and average endurance of 10 minutes.

The Snipe continued in production in 1977, when about 30 were in use, most of them with overseas operators.

(France)

AÉROSPATIALE CT.20

Target drone and tug for towed target; first flight late 1950s
Power plant: One Turboméca Marboré II turbojet engine (880 lb; 400 kg st) or one Marboré VI (1,056 lb; 480 kg st)
Wing span: 10 ft 4½ in (3.16 m)
Length overall: 17 ft 10½ in (5.45 m) standard; 18 ft 4½ in (5.60 m) for extended version with trailed target
Body diameter (max): 2 ft 2 in (0.66 m)
Wing area: 33.34 sq ft (3.20 m^2)
Weight empty: 1,080 lb (490 kg) standard; 1,344 lb (610 kg) version with trailed target
Max launching weight: 1,455 lb (660 kg) standard; 1,763 lb (800 kg) version with trailed target
Max speed at 32,800 ft (10,000 m):
Marboré II 485 knots (560 mph; 900 km/h)
Marbore VI 512 knots (590 mph; 950 km/h)
Max operating height (standard version): 45,925 ft (14,000 m)
Min operating height: 100 ft (30 m)
Time to 32,800 ft (10,000 m): 6 min
Endurance (standard version):
at max operating height 50 min
at min operating height 15 min
Practical range of guidance/tracking system: 135 nm (155 miles; 250 km)

Launch and recovery: Launched by booster rockets from vehicle-mounted ramp, as illustrated. Recovery by parachute and impact-absorbing airbag deployed after engine is stopped by transmission of landing signal. Can be recovered from land or water.
Guidance and control: Radio command guidance system, operated by controller on ground or in airborne director aircraft. Nine basic signals can be transmitted: turn right; turn left; nose up; nose down; increase power; reduce power; trace smoke; operate cameras; and land.
Special equipment: Can be used to tow a Dornier target which is pylon-mounted under starboard wing at launch; or can be used with trailed target system, primarily for training in the use of air-to-air missiles with electro-magnetic (BEY) or infra-red homing (EMIR) systems. Version known as CT.20TBA has a TRT AVH-6 radio altimeter, for very low (about 100 ft; 30 m) altitude capability, and an improved remote guidance system.
Remarks: In production since 1958; by September 1976 a total of 1,221 had been built, including 281 for export to NATO and non-NATO countries such as Italy and Sweden. Variants include R.20 reconnaissance drone, described separately.

CT.20 with underwing towed target

AÉROSPATIALE C.22 (France)

Variable-speed target drone
Power plant: One Microturbo TRI-60 turbojet engine (772 lb; 350 kg st); one solid-propellant rocket booster (8,157 lb; 3,700 kg st)
Wing span: 8 ft 2½ in (2.50 m)
Length overall: 16 ft 8 in (5.10 m)
Span of tail-fins: 3 ft 11 in (1.20 m)
Height overall (incl towed target): 3 ft 9 in (1.15 m)
Weight of booster: 143 lb (65 kg)
Max launching weight (incl booster): 1,168 lb (530 kg)
Max level speed: Mach 0.95
Operating height range: 50 to 45,930 ft (15 to 14,000 m)
Max endurance at Mach 0.9 at 42,650 ft (13,000 m): 1 hr 30 min

Launch and recovery: Launched by rocket booster from mobile launcher. Parachute recovery. Inflatable airbags to assist in water recovery or to cushion ground impact.
Guidance and control: Radio command guidance system.
Special equipment: Radio command link equipment and radio altimeter. Underfuselage attachments for a 5 ft 7 in (1.70 m) long tow "bird" target, weighing approx 55 lb (25 kg), which can be towed on a cable up to 2,625 ft (800 m) behind the C.22. This target also available separately. Can be fitted with Luneberg lenses for radar augmentation or propane heat source for infra-red augmentation.
Remarks: Intended successor to CT.20 (which see), for service in the 1980s; announced in late 1976. Capable of re-use for up to 12 missions.

Left: *C.22 with target 'bird' under fuselage.* Right: *D 15 low-cost twin-engined target drone*

AÉROSPATIALE D 15

Experimental target drone; first flight 10 April 1975
Power plant: Two 1.7 hp piston engines
Wing span: 7 ft 10½ in (2.40 m)
Length overall: 7 ft 2½ in (2.20 m)
Wing area: 10.76 sq ft (1.00 m^2)
Payload: 13.2 lb (6 kg)
T-O weight: 39.7 lb (18 kg)

Performance: Can simulate a low-level strike at speeds of up to 108 knots (124 mph; 200 km/h) or diving attacks at up to 162 knots (186 mph; 300 km/h)
Endurance: 1 hr at full power
Guidance and control: Radio command guidance.
Remarks: Developed as small, inexpensive target drone for anti-aircraft artillery training. Construction mostly of plastics materials.

(France) AÉROSPATIALE R.20

Battlefield reconnaissance drone; first flight 1963
Power plant: One Turboméca Marboré IID turbojet engine (880 lb; 400 kg st); two solid-propellant booster rockets for launch
Wing span: 12 ft 2½ in (3.72 m)
Width, wings folded: 4 ft 5 in (1.35 m)
Length overall: 18 ft 9 in (5.71 m)
Body diameter (max): 2 ft 2 in (0.66 m)
Payload: 330 lb (150 kg)
Launch weight:

excl booster	1,875 lb (850 kg)
incl booster	2,425 lb (1,100 kg)

Operating speed: Mach 0.65
Operating height range: 660-32,800 ft (200-10,000 m)
Operating radius at low altitude: 86 nm (100 miles; 160 km)
Launch and recovery: Launched by two booster rockets from short ramp on standard Berliet GB-C8-KT French Army lorry. Recovery system similar to that of CT.20.
Guidance and control: Radio command guidance system. When close to launch post, the R.20 is controlled directly from the ground. Over longer distances it is controlled automatically by a gyroscopic platform and an electronic programmer, enabling it to follow a prearranged flight plan.
Special equipment: Radio link for sending back data during flight. Three synchronised Omera 114 x 114 mm cameras, which can photograph more than 77 square miles (200 km^2) during a single low-altitude sortie. Flares can be carried for night photography, or SAT Cyclope infra-red line-scanning equipment.
Remarks: Developed from Aérospatiale CT.20 target drone (which see); in production for French Army. Total of 62 built by September 1976, with production continuing.

R.20 battlefield reconnaissance drone on its launch truck

AMR RPV-004 and AMR-RYAN RPV-007

Experimental mini-RPVs
Data apply to RPV-007
Power plant: One K & B piston engine (2.2 hp), driving a four-blade shrouded propeller (Rotorduct)
Wing span: 5 ft 0 in (1.52 m)
Max T-O weight: 12 lb (5.4 kg)
Launch and recovery: Conventional take-off and landing.
Guidance and control: Radio command guidance system. Aerodynamic control by differentially-operating elevons and twin rudders.
Special equipment: Autopilot.

AMR RPV-004 (left) and RPV-007 mini-RPVs. The electronics bay of the 004 is located beneath the striped hatch in the centre of the wing. The 007 has a dorsally mounted Rotorduct propulsion system

Variants:
RPV-004. Earlier model than RPV-007, for which it served as systems testbed. Has also been demonstrated to US Army and Navy; has air-tested autopilot systems; and has taken off under ground remote control, been handed over to helicopter airborne control while in flight, and returned to ground control for landing. Max speed 113 knots (130 mph; 209 km/h); max rate of climb at S/L 1,200 ft (366 m)/min. Aerodynamic control by elevons and central elevator; no rudders.
RPV-007. Designed, built (several examples) and flown as half-scale testbed for Teledyne Ryan Model 262 (which see). Radio control and autopilot were first flight-tested in the RPV-004.

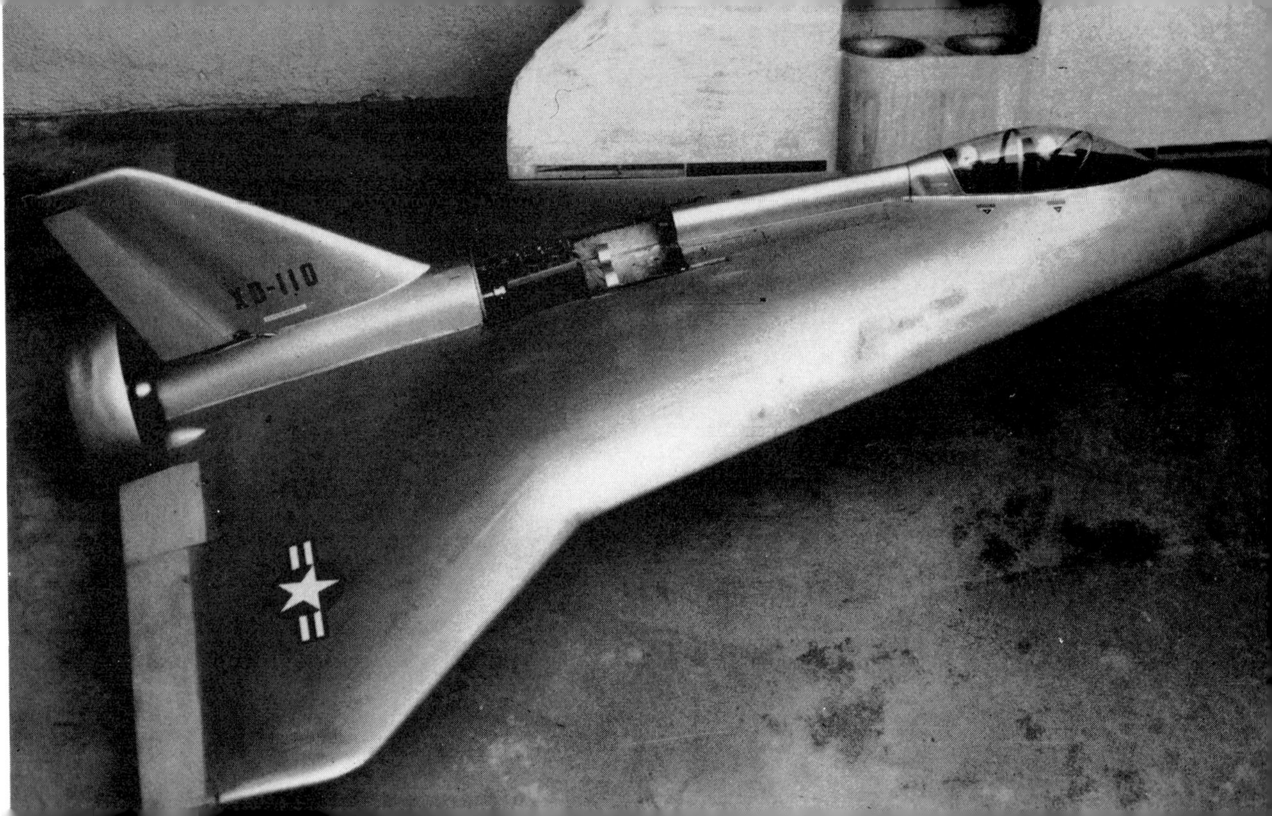
XD-110

AMR XD 110

Experimental mini-RPV; first flight 1969
Power plant: One Ross six-cylinder piston engine (4 hp), driving a multi-blade shrouded propeller (Rotorduct)
Wing span:
XD 110A 6 ft 6 in (1.98 m)
XD 110B 7 ft 0 in (2.13 m)
Wing chord at centreline:
XD 110A 9 ft 3 in (2.82 m)
XD 110B 10 ft 0 in (3.05 m)
Max payload:
XD 110A 50 lb (22.7 kg)
XD 110B 80 lb (36.3 kg)
Max level speed:
XD 110A 228.2 knots (262.8 mph; 422.9 km/h)
XD 110B 260 knots (300 mph; 482 km/h)
Cruising speed:
XD 110A 139 knots (160 mph; 257 km/h)
Min flying speed:
XD 110A 11.03 knots (12.7 mph; 20.4 km/h)
XD 110B 9.03 knots (10.4 mph; 16.7 km/h)
Endurance: more than 12 hr
Launch and recovery: Conventional T-O and landing, using retractable tricycle landing gear. Optional net recovery system for XD 110B.
Guidance and control: Radio command guidance system. Aerodynamic control by elevons and rudder.
Special equipment: XD 110A has been used as testbed for autopilot, laser sensor and TV navigation systems. XD 110B scheduled to air-test electronics packages and a five-axis autopilot.
Variants:
XD 110A. Initial version, third prototype of which was serving in 1976 as scale test model for a two-seat STOL advanced trainer. Non-stalling double-delta wing with sealed control surfaces.
XD 110B. Designation of fourth prototype. Slightly larger than XD 110A, with twin Rotorducts buried in rear fuselage. Undergoing flight testing in 1976.

AMR XD 110A research scale model for the two-seat Benson Nova training aircraft

NAVY
RPD2

(USA)

APL RPD2

Mini-target for naval fire control systems evaluation; first flight 13 June 1975
Power plant: One McCulloch MC-101A single-cylinder piston engine (12 hp), driving a two-blade pusher propeller
Dimensions: No details known
Payload: 22 lb (10 kg)
Max launching weight: 80 lb (36.3 kg)
Max level speed: 150 knots (173 mph; 278 km/h)
Cruising speed: 65 knots (75 mph; 120 km/h)
Service ceiling: 18,000 ft (5,500 m)
Endurance: 1 hr 30 min

Launch and recovery: Launch from small trolley, by rubber cord, or from larger catapult system. Parachute recovery system with airbag shock-absorbers.
Guidance and control: System includes autopilot which uses Earth's electrostatic field for pitch and roll stabilisation.
Special equipment: Includes passive radar augmentation devices housed in nose and in fairing on fin trailing-edge.
Remarks: Developed from an earlier US Army programme vehicle by the Applied Physics Laboratory of Johns Hopkins University. Under test in 1975 by US Naval Surface Weapons Center as target for improved shipboard fire control systems.

APL RPD2 mini-RPV target for shipboard operation

AJEP
TAILWIND
G-AYDU

BAC/AJEP (WITTMAN) TAILWIND SERIES 2

Target drone; first flight under remote control late 1976
Data apply to piloted version; drone version similar
Power plant: One Continental PC60 flat-four piston engine (100 hp)
Wing span: 22 ft 6 in (6.86 m)
Length overall: 20 ft 10½ in (6.36 m)
Wing area: 90.0 sq ft (8.36 m^2)
Max T-O and landing weight: 1,400 lb (635 kg)
Max level speed at S/L: 152 knots (175 mph; 282 km/h)
Max cruising speed at S/L: 143 knots (165 mph; 265 km/h)
Range with max payload, no reserves: 534 nm (615 miles; 990 km)
Launch and recovery: Conventional T-O and landing, using non-retractable tricycle landing gear.
Guidance and control: Radio command guidance system. Aerodynamic control by conventional wing and tail control surfaces. Manual controls retained, to permit normal piloted operation when not required for use as a drone.
Special equipment: Autopilot and radio command link equipment.
Remarks: Under development by BAC's Guided Weapons Division, the Tailwind 2 is intended to provide missile operators with a realistic full-size target vehicle, and may become an integral part of the company's Rapier air defence system. Prototype development began in 1976, using the AJEP prototype Tailwind Series 2, a British development of the original US Wittman design.

AJEP Tailwind Series 2 prototype, converted by BAC for unmanned operation

(USA) BEECHCRAFT MODELS 1001 (MQM-39A) and 1025 (MQM-61A CARDINAL)

Target drone; first flight 1958 (?)
Power plant: One McCulloch TC6150-J-2 six-cylinder piston engine (125 hp); provision for JATO bottle under rear fuselage for zero-length launch
Wing span: 12 ft 11½ in (3.95 m) over wingtip containers
Length overall: 15 ft 1 in (4.60 m)
Body diameter: 1 ft 5 in (0.45 m)
Wing area: 24.4 sq ft (2.27 m^2)
Flight weight: 560 lb (254 kg)
Max launching weight with JATO: 664 lb (301 kg)
Max level speed: over 303 knots (350 mph; 560 km/h)
Service ceiling: above 43,000 ft (13,100 m)
Endurance: more than 1 hr at 25,000 ft (7,620 m)
Launch and recovery: Zero launch from take-off trolley, using JATO booster. Parachute recovery.
Guidance and control: Radio command guidance system. Aerodynamic control by ailerons and elevator/rudders.
Special equipment: Miss-distance and scoring systems to evaluate accuracy of anti-aircraft missiles and gunfire. Molten infra-red sources carried when testing heat-seeking missiles. Radar augmentation devices can also be carried. Cardinal used also for towing targets (two carried under each wing) and banners.
Variants:
MQM-39A (formerly KDB-1). Winner of 1955 US Navy design competition. In production 1959-71, during which time approx 800 built for US Navy, Marine Corps and Army, and for Germany and Switzerland. Used as out-of-sight target for air-to-air and surface-to-air missiles, rockets and predicted-fire weapons.
MQM-61A Cardinal. Major version, of which 20 built for Spanish Army and more than 1,300 for US Army before production ended in Spring 1972. Used to train crews of Hawk, Redeye, Chaparral and other missiles.

Beechcraft Model 1001 target supplied to Germany

AQM-37A
AQ-11641

BEECHCRAFT MODELS 1019 and 1102 (AQM-37A)

Expendable air-launched target drone; first flight 31 May 1961
Power plant: One Rocketdyne/AMF LR64 P-4 twin-chamber liquid-fuel rocket motor (631 lb; 286 kg st)
Wing span: 3 ft 3½ in (1.00 m)
Length Overall: 12 ft 6½ in (3.82 m)
Body diameter: 1 ft 1 in (0.33 m)
Height overall: 1 ft 8 in (0.51 m)
Wing area (exposed): 9.35 sq ft (0.87 m^2)
Max launching weight: 565 lb (256 kg)
Operating speed range: Mach 0.4 to Mach 3.0
Operating height range: 1,000 to 80,000 ft (300 to 24,385 m)
Max range: more than 100 nm (115 miles; 185 km)
Max endurance: 15 min
Launch and recovery: Air-launched from launcher which is adaptable to various fighter aircraft. Non-recoverable.
Guidance and control: Programmed radio guidance system. Aerodynamic control by full-span ailerons and canard foreplanes. Flight normally terminated by aerodynamic means, but explosive destructor system is available to provide additional range safety and operational flexibility.
Special equipment: Target provides both active and passive radar augmentation for radar acquisition and lock-on. Chemical flare for missions requiring infra-red augmentation. Choice of two miss-distance scoring systems.
Variants:
Model 1019 (AQM-37A). For aircraft and missile threat simulation, defence weapon system evaluation and operational crew training. Operational with US Navy since 1963; current launch aircraft are A-4, F-4 and F-8. Modified examples evaluated in 1968 by USAF (10) and US Army (3). Further 10 due for delivery to US Army in 1976, followed by 28 of a parachute- recoverable AQM-37A which will be built in high-altitude and low-altitude versions. Total production (all versions, including export models) has exceeded 3,000.
Sea Skipper. US Navy programme (originally Sea Skimmer) under which, from late 1974, Beech modified 10 AQM-37As for low-altitude flight missions, capable of being air-launched over the sea at 1,200 ft (366 m) and then flown as low as 50 ft (15 m), to evaluate ship defence systems. Entered service 1975. Delivery of further 10 began in Spring 1976.
Models 1072, 1088, 1094 and 1095. Export versions, described separately.
Model 1102. Improved AQM-37A with modified wings, solid-state autopilot and other changes; 100 ordered by US Army in early 1977.

AQM 37A air launched target on its transport trolley

FLIGHT

(USA)

BEECHCRAFT MODELS 1072, 1088, 1094 and 1095

Export versions of Model 1019 (see previous entry)
Power plant, Dimensions, Weights and Performance: Generally similar to Model 1019 (which see)
Variants:
Model 1072. Version for UK, substantially re-engineered by Beech and Short Bros & Harland as the **SD.2 Stiletto** to meet British requirements, including virtually complete replacement of the radio and radar systems and control system changes. Principal modifications are incorporation of an EMI T44/1 telemetry system; provision of additional 15V flight breakup system (WREBUS); installation of radioactive miss-distance indicator (RAMDI), with associated radio link; introduction of Plessey IR 112A/IR 310 telecommand system and heading and turns command circuitry; modification to propulsion system to give Mach 2 performance at 60,000 ft (18,300 m); and changes in the radar augmentation system. Various optional mission kits, which can be installed either by Shorts or by operator. First Stiletto launched at Llanbedr on 1 August 1968; last of 75 ordered due for delivery January 1976. Launch aircraft is Canberra PR Mk 3.
Model 1088. Manufacturer's designation of five supplied to Italy. For air-launch from F-104S aircraft of Italian Air Force.
Model 1094 (Matra Vanneau: Lapwing). Fifteen Model 1094s ordered in mid-1973 by French Air Force, of which deliveries began about a year later. A follow-on contract for 30 was announced in August 1974; these are being modified to French Air Force requirements by Matra.
Model 1095. Ten Model 1095s ordered in August 1974 by British Ministry of Defence. These are being modified by Shorts to MoD specifications and are for use in crew training exercises on the Hebrides range.

Model 1094 Vanneau, modified by Matra for the French Air Force

TARGET
U.S. AIR FORCE

(USA) BEECHCRAFT MODEL 1070 HAST

Air-launched target drone; first flight 9 July 1975
Power plant: One CSD hybrid rocket engine (thrust variable from 120 to 1,200 lb; 54 to 544 kg)
Wing span: 3 ft 4 in (1.02 m)
Length overall: 16 ft 8 in (5.08 m)
Body diameter: 1 ft 1 in (0.33 m)
Wing area (total exposed): 10.44 sq ft (0.97 m^2)
Payload: 85 lb (38 kg)
Launching weight: 1,145 lb (519 kg)
Operating speed range: subsonic to Mach 4.0 (460 to 2,345 knots; 530 to 2,700 mph; 853 to 4,345 km/h)
Operating height range: 40,000 ft (12,200 m) to 100,000 ft (30,500 m)
Endurance at Mach 3: 5 min
Launch and recovery: Air-launched from carrier aircraft (initially, F-101, F-4, F-14 and F-15), at speeds between Mach 0.8 and 2.5. Recovery by 45 ft (13.72 m) parachute from land, water or mid-air.
Guidance and control. Manoeuvres can either be pre-programmed or initiated via ground command radio link. Aerodynamic control by full-span ailerons and canard foreplanes. Manoeuvres of between 5*g* at 35,000 ft (10,670 m) and 1.15*g* at 90,000 ft (27,400 m) are part of flight test programme; vehicle is also to be capable of "S" and 180° turns in the horizontal plane and altitude changes in the vertical plane.
Special equipment: Modular payloads with a wide variety of options will be available for accurate simulation of aircraft or missile threats, and will include various radar and infra-red augmentation devices as well as a flare/chaff dispenser. Vector miss-distance scoring systems will also be included.
Variants:
Sandpiper. USAF experimental programme of 1967-68 for a high-performance rocket-powered target drone. Programme confined to verifying the hybrid propulsion system and command manoeuvring system, using modified AQM-37A drones as flight test vehicles. Phase 2 superseded by HAST programme.
Model 1070 HAST (High Altitude Supersonic Target). Current USAF programme, on behalf of all three US services. Twelve flight test vehicles and 13 refurbishment kits built under 1971 contract; delivery of the former began in 1972, followed by a captive flight programme and, in August 1974, a contract for a 28-month flight test programme. The latter was continuing in 1976.

Model of the Beechcraft 1070 High Altitude Supersonic Target (HAST)

ARMY
Beechcraft
VSTT
KD-8
MODEL 1089

(USA)

BEECHCRAFT MODEL 1089 (MQM-107 STREAKER)

Target and target-towing drone; first flight 1973 (?)
Power Plant: One Teledyne CAE 372-2 (J402-CA-700) turbojet engine (640 lb; 290 kg st), and one JATO booster rocket
Wing span: 9 ft 10 in (3.00 m)
Length overall: 16 ft 10 in (5.13 m)
Body diameter: 1 ft 3 in (0.38 m)
Height overall: 4 ft 10 in (1.47 m)
Wing area (total projected): 27.16 sq ft (2.52 m^2)
Max launching weight (incl booster): 1,014 lb (460 kg)
Operating speed range: 247 to more than 500 knots (285-576 mph; 459-927 km/h)
Operating height range: S/L to 40,000 ft (12,200 m)
Endurance: more than 3 hr
Launch and recovery: Lightweight launcher and checkout system, easily transportable in two suitcase-size containers. Parachute recovery.
Guidance and control: Rado command guidance of manoeuvre and recovery, which can be either pre-programmed or maintained directly during flight. In flight, guidance and control system automatically stabilises about roll, yaw and pitch attitudes, and provides altitude and speed hold modes. Aerodynamic control by ailerons and elevators.
Special equipment: As target, can carry appropriate equipment for testing such air defence systems as Chaparral, Redeye, Hawk and Stinger. As target tug, is able to carry various tow targets underwing (eg, two TA-8 radar augmentation or infra-red augmentation targets) on each mission; these can be towed separately up to 8,000 ft (2,440 m) behind the drone.
Remarks: Winner in 1975 of US Army Missile Command's VSTT (Variable-Speed Training Target) competition, and expected to be that service's primary subsonic missile training target. Initial production contract announced 4 April 1975; deliveries began in Spring 1976. Eventual US Army requirement for 317 is expected, of which 176 ordered by Autumn 1976. Modified version (Model 1089E) is a competitor in US Air Force TEDS competition (which see).

MQM-107 Streaker variable-speed training target

USAF

BOEING YQM-94A B-GULL (COMPASS COPE B)

High-altitude long-endurance strategic RPV; first flight 28 July 1973
Power plant: One General Electric J97-GE-100 turbojet engine (5,270 lb; 2,390 kg st)
Wing span: 90 ft 0 in (27.43 m)
Length overall (excl nose probe): 40 ft 0 in (12.19 m)
Wheel track: 21 ft 0 in (6.40 m)
Weight empty: 5,500 lb (2,494 kg)
Payload for 24 hr mission: 700 lb (317 kg)
Max T-O weight: 14,400 lb (6,531 kg)
Cruising speed at 50,000-70,000 ft (15,240-21,340 m): Mach 0.5 to 0.6
Launch and recovery: Conventional take-off and landing, using retractable tricycle landing gear. Provision for all-weather automatic landing system.
Guidance and control: Ground control via command module which embodies standard cockpit instrumentation, TV screen and navigation display; data link equipment includes microwave command digital guidance system; wide-band microwave data transmission system for return of video signals; and TPW-2A X-band radar van. Sperry integrated analogue flight control system, with internally-generated ILS; redundant stabilisation system; APW-26 airborne transceiver; and other data link equipment. TV camera in undernose fairing. Aerodynamic control by wing airbrakes and ailerons, elevator and twin rudders.
Special equipment: Except for TV camera, details of operational equipment are classified. All sensors and antennae housed in lower half of front fuselage.
Remarks: Winner in 1976 of US Air Force Compass Cope programme to select a high-altitude, long-endurance RPV to replace such manned reconnaissance aircraft as RB-57 and RC-135 for use in sensitive areas. Two prototypes built for 1973-75 evaluation, the second of which, on 23/24 November 1974, completed an endurance trial of 17 hr 24 min and attained altitudes of more than 55,000 ft (16,765 m). Pre-production version (three ordered) will have a 7,500 lb (3,402 kg) st General Electric turbofan engine, digital flight control system, Bendix automatic landing system, and Sperry Univac continuous-wave frequency-modulated command guidance system. Payload capacity will be increased to 700-750 lb (317-340 kg). US Navy has expressed interest in similar vehicle for ocean surveillance, possibly carrier-based. Also under consideration by USAF for Precision Location Strike System (PLSS) role, acting as relay for strike aircraft.

Left: *Artist's impression of production-type Compass Cope B, showing nose TV camera.* Right: *Second prototype of the YQM-94A*

USAF

BRUNSWICK MAXI-DECOY and PROPELLED DECOY

Decoy drones; first flight 1974
Data: A: Maxi-Decoy; B: Propelled Decoy
Power Plant (B): One Atlantic Research solid-fuel rocket motor, burn time 5 min
Wing span:
A 2 ft 6 in (0.76 m)
B 4 ft 7 in (1.40 m)
Body:
A (width) 5 in (0.13 m)
B (diameter) 10 in (0.25 m)
Length:
A 3 ft 9 in (1.14 m)
B 7 ft 4 in (2.24 m)
Payload volume:
A 500 cu in (0.008 m^3)
B 1,800 cu in (0.03 m^3)
Launch weight:
A 130 lb (59 kg) (max)
B 300 lb (136 kg)
Max speed:
A, B Mach 0.8 to 0.9
Remarks: Developed for carriage by USAF strike aircraft, on whose external stores points they are carried. Equivalent to an "iron bomb", the decoys are carried with wings folded on the aircraft external stores points. Shortly after release the wings are opened by internal command to commence flight. Flight profiles and manoeuvres are pre-programmed.
Variants:
Maxi-Decoy. Unpowered decoy glider with square-section body. Under development primary as active or passive ECM carrier. ECM payloads, developed by the Automated Systems Division of RCA, include C-band jammer; or G-band jammer combined with radar signature augmentation. Payload test flights 1977, with F-4 carrier aircraft. Three Maxis can be accommodated in each of the F-4's four underfuselage Sparrow missile recesses, and two on each 750 lb (340 kg) underwing hardpoint; several Maxis can be deployed simultaneously.
Propelled Decoy. Larger than Maxi, with circular body and cruciform sweptback tail surfaces. Power plant permits it to cruise at high subsonic speeds, and two-axis autopilot allows it to manoeuvre at high cruising speeds as well as performing in glide modes. Also intended primarily for carrying active and passive ECM. One Propelled Decoy can be carried on each of the F-4's 750 lb (340 kg) underwing hardpoints.

Maxi-Decoy unpowered decoy, with flip-out wings

(Canada) CANADAIR CL-89

Battlefield surveillance RPV; first flight 1964
Power plant: One Williams Research WR2-6 turbojet engine (125 lb; 56.7 kg st); one Bristol Aerojet Wagtail rocket booster motor (4,550 lb; 2,065 kg average thrust)
Wing span: 3 ft 1 in (0.94 m)
Body diameter: 1 ft 1 in (0.33 m)
Length overall, excl nose probe: 12 ft 3 in (3.73 m) incl booster; 8 ft 6½ in (2.60 m) excl booster
Payload: 33.3 lb (15.1 kg)
Max launching weight: 343 lb (156 kg) incl booster; 238 lb (108 kg) excl booster
Max speed: 400 knots (460 mph; 741 km/h)
Launch and recovery: Launched from truck-mounted short ramp; booster separates automatically after 2½ sec of flight. For recovery, after final positioning by the ground homing beacon, the drone is slowed by a drogue parachute until the main parachute deploys. The drone is then inverted, and forward and rear airbags are inflated and deployed automatically to absorb landing impact.
Guidance and control: Flight path, altitude and sensor on/off commands are controlled by a preset programmer which receives information from an onboard Air Distance Measuring Unit (ADMU) and combines this with the preset programme to control the flight path. A ground homing beacon positions the drone in its final stages of flight to ensure accuracy of landing.
Special equipment: Two main sensor systems in current use: Carl Zeiss KRb 8/24 camera system, and Hawker Siddeley Dynamics Type 201 infra-red linescan (IRLS) system. Other onboard equipment includes recovery parachutes and landing airbags; ADMU; programmer; static power converter; homing receiver; amplifier; flash detector; directional and vertical gyros; transponder antenna; air bottle; and a dorsal pack containing 12 photoflares.
Remarks: Development began in 1961 as a joint Canadair/Canadian government venture, and production began in 1967. The system, which has the NATO designation **AN/USD-501**, has since been produced for Canada, the Federal Republic of Germany, Italy (jointly with Meteor SpA) and the UK. More than 500 CL-89 drones produced by 1976, with production continuing. A scaled-up derivative designated **CL-289** (AN/USD-502) is under joint development by Canadair and Dornier to meet a German requirement for a longer-range vehicle (approx 215 nm; 248 miles; 400 km). It will also have twice the payload capacity, photographic and IRLS sensors, and a real-time data transmission system for IRLS imagery.

AN/USD-501 (CL-89) surveillance RPV leaving its launch vehicle

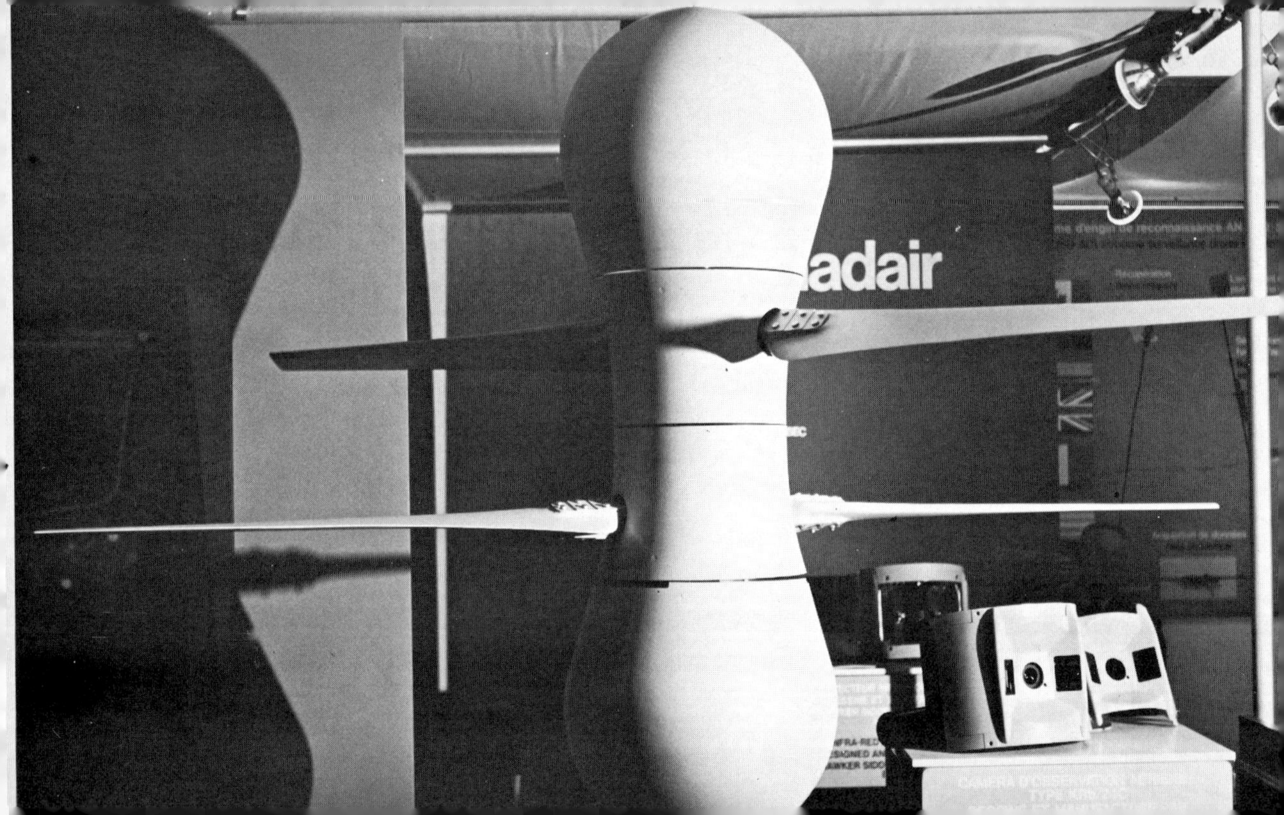
adair

(Canada)

CANADAIR CL-227

Battlefield surveillance/target acquisition/fire adjustment RPV
Rotor diameter (each): 7 ft 6 in (2.29 m)
Length of body: 5 ft 0 in (1.52 m)
T-O weight: 200 lb (91 kg)
First shown in model form at the 1975 Paris Air Show, the CL-227 was designed to meet a British Ministry of Defence requirement. It has an "hour-glass" shaped body, the power plant occupying the upper bulged portion and driving two co-axial contra-rotating rotors amidships to provide VTOL capability. A small simple platform will be used for vertical take-off and landing.

The lower bulged portion of the vehicle houses the payload, which can include various types of electro-optical sensor such as TV, low light level TV and thermal imaging systems. The CL-227 can handle real-time data transmission and has a radio command guidance system.

Model of the Canadair CL-227 (Brian M. Service)

DORAND DS.7

Remotely controlled gyro-glider; first flight 1973 (?)
Power plant: None
Rotor: Two-blade telescopic type, with max diameter of 13 ft 1½ in (4.00 m). Each blade made in two sections, the outer portion of 7.9 in (20 cm) chord telescoping into the 11.8 in (30 cm) inboard portion
Suspension: Three struts unfold and lock after deployment to form a 1 ft 11½ in (0.60 m) pylon, attached at one end to the rotor hub and at the other to the payload container.
Payload: 485 lb (220 kg)
Gross weight incl payload: 551 lb (250 kg)
Control system: Comprises cyclic control swashplate, cam-operated programmer, and linkages between cam followers and rotor blades. Programmer cams started by onboard height detector and driven by rotor to generate control movements necessary to change collective and cyclic pitch angles in accordance with a flare-out programme at touchdown. Fantail stabiliser fitted, to keep payload container pointed into wind.
Remarks: Developed under DRME contract as a potential method of soft-landing and recovering, under pre-programmed control, loads dropped from aircraft by providing the load with the means of autorotative descent. The DS.7, under evaluation in 1974-75, is a half-scale test vehicle. Applications foreseen, if tests are successful, include recovery of loads at very low altitudes or in high wind conditions; recovery of Aérospatiale R.20 reconnaissance drones (which see); steering of a dropped load to a predetermined point under remote control; automatic recovery of a helicopter following an engine failure; ejection seat recovery; and as a personnel gyro-glider. Payloads of up to 33,070 lb (15,000 kg) could be recovered with a system of this type, depending upon rotor size, and release altitude could be as low as 50 ft (15 m).

Dorand DS.7 half-scale gyro-glider test model

SYSTEM

(Germany)

DORNIER AERODYNE E 1

VTOL experimental RPV; first flight (tethered) 18 September 1972
Power plant: One MTU 6022A-3 turboshaft engine (370 shp), driving a five-blade shrouded fan propeller; replaced in 1976 by Avco Lycoming LTS 101-650C turboshaft of 592 shp
Length overall: approx 18 ft 0½ in (5.50 m)
Max diameter: approx 6 ft 2 in (1.90 m)
Fan diameter: approx 3 ft 7 in (1.10 m)
Max T-O weight: 959 lb (435 kg)
T-O and landing: VTOL or V/STOL
Guidance and control: Vertical lift by deflected thrust from fan shroud via deflection chamber and non-moving cascade vanes; pitch and yaw during hovering controlled by regulating engine exhaust via tailboom; conventional tail surfaces would be fitted to control pitch and yaw in forward flight. Keel flap for roll control during hovering and forward flight. Dornier radio command guidance system.
Special equipment: Bodenseewerk onboard attitude control system, incorporating a vertical gyro, three rate gyros, and three accelerometers. Telemetry data link system, with more than 50 channels. Power supply system incorporates battery for emergency power in the event of a generator failure.
Remarks: Development, begun by Dr A.M. Lippisch in the USA, has been continued by Dornier since 1967 under German Ministry of Defence contract. Initial hovering test programme, comprising 74 flights totalling almost 1½ hr in the air, ended on 30 November 1972. Test flights from moving platform scheduled for Autumn 1976, for which tail unit, reinforced landing gear and LTS 101-650C engine being fitted. Dornier and Hawker Siddeley Dynamics (UK) have, under a 1971 agreement, developed a wingless VTOL maritime RPV (see next entry) based on the Aerodyne concept, and operational definition studies in 1974 confirmed the Aerodyne as a suitable V/STOL RPV for various naval applications. Proposals for such applications have been submitted to NATO.

Aerodyne E 1 prototype VTOL RPV in tethered flight

0 1 2 3 4 5 6 7 8 9 10

DORNIER/HSD MARITIME AERODYNE

Wingless VTOL RPV project
Power plant: One turboshaft engine, driving two nine-blade shrouded fan propellers with cascade-type thrust vector control vanes in rear of each duct
Width overall: 9 ft 10 in (3.00 m)
Length overall: 16 ft 9 in (4.90 m)
Height overall: 8 ft 6 in (2.60 m)
Max T-O weight: 3,000 lb (1,360 kg)
Max speed: 329 knots (379 mph; 610 km/h)
Ceiling for horizontal flight: 22,000 ft (6,700 m)
Hovering ceiling: 3,940 ft (1,200 m)
Range at max speed: 809 nm (932 miles; 1,500 km) with 220.5 lb (100 kg) payload
Endurance:
with 882 lb (400 kg) payload 1 hr 20 min
with 220.5 lb (100 kg) payload 3 hr 20 min
Launch and recovery: Conventional T-O and landing using retractable three-wheel landing gear (one wheel in underside of each fan duct and one below underfin).
Guidance and control: Flight stabilisation and control in all three axes by adjusting vectored thrust and (at high speeds) T tailplane.
Special equipment: Depending upon mission, may include equipment for surveillance; over-the-horizon fire control; electronic countermeasures; anti-submarine warfare; weapon delivery (eg, two underslung torpedos); or communications/intelligence relay. Suitable for use from small ship decks.

Left: *Model of the Maritime Aerodyne.* Right: *Artist's impression of the ASW version of the Aerodyne, with underfuselage torpedos*

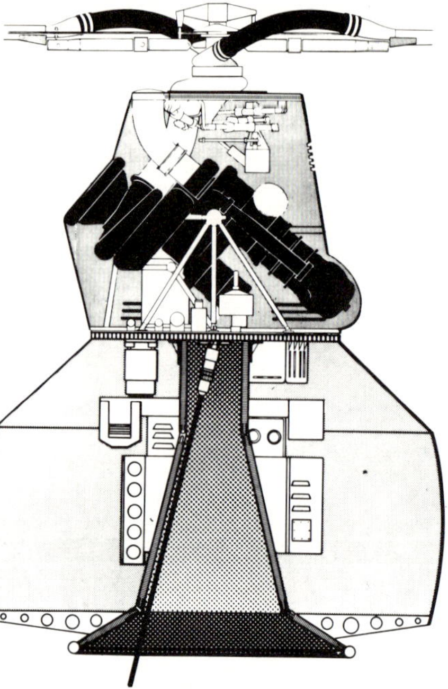

DORNIER Do 32, Do 34 KIEBITZ and ARGUS

Multi-purpose tethered observation platform; first flight (Do 34) due 1977
Data apply to Do 34 Operational Kiebitz
Power plant: One Allison 250-C20 turboshaft engine (400 shp rating), providing cold-jet air via MTU radial compressor to blade tips to drive rotor
Rotor diameter: 26 ft 3 in (8.00 m)
Body diameter at base: 3 ft 5 in (1.05 m)
Height overall: 4 ft 9 in (1.45 m)
Weight without tether or payload: 617 lb (280 kg)
Max payload to 985 ft (300 m), ISA: 308 lb (140 kg)
Reel-in/reel-out speed: 9.85 ft (3 m)/sec
Operational ceiling: 985 ft (300 m)
Mission endurance: 24 hr
Launch and recovery: Not applicable. Kiebitz is a cable-tethered RPV, housed in a vehicle which incorporates its landing platform, winch, control post, fuel tank and auxiliary equipment.
Guidance and control: After arrival on site, can be in position at operational height of 985 ft (300 m) in 8 minutes; limiting factors are a wind speed of 46 ft (14 m)/sec ± 26 ft (8 m)/sec, available thrust reserves, and sensor requirements. Control system aligns drone according to airframe attitude and position in relation to ground. An electromagnetic sensor measures any drift from desired position.
Special equipment: Appropriate payload packages for reconnaissance, fire control, communications or traffic monitoring duties. Cone-shaped airframe is designed to minimise radar reflectivity, and payload compartment is located in base of this, enabling quick change of sensors and providing space for a large-volume radome. One prototype has Decca RDL-2 passive ECM detector, operating in 0.5-18CHz band; other (Argus) prototype has modified LCT Orphée radar. Naval version, with Ferranti Seaspray search radar, has been proposed.
Variants:
Do 32 U. Drone version of Do 32 E single-seat ultra-light foldable helicopter, first flown 1966. Three-axis flight control system. Drive for cold-jet rotor from MAN-Turbo 6012L turbocompressor. Rotor diameter 24 ft 7 in (7.50 m), max T-O weight 705 lb (320 kg).
Do 32 K Experimental Kiebitz. Feasibility version of Kiebitz, utilising Do 32 U engine, rotor, transmission and flight control systems. Later re-engined with KHD T 212. First flown late 1970. Five built, which had completed approx 100 hr of tethered test flying by 1974, at heights of up to 660 ft (200 m). Experimental programme still active in 1975.
Do 34 Operational Kiebitz. Operational version, as described. Two prototypes and two ground stations ordered in August 1972. Due to begin flight test in 1977.
Argus (Autonomes Radar Gefechtsfeld Uberwachungs System). Joint German-French programme for battlefield surveillance system, being developed under 1974 agreement. Consists of a Do 34 Kiebitz prototype, fitted with advanced version of LCT Orphée radar as primary sensor, mobile ground station, and tether cable. Due to begin flight test in 1977.
Seekiebitz. Projected maritime version with Ferranti Seaspray search radar.

Far left: *Dornier Experimental Kiebitz hovering over its mobile ground control station.* Left: *Drawing of Argus. Engine is shown in solid black, radar location indicated by central shaded area*

DORNIER UKF

(Germany)

Experimental pilotless aircraft; under development

The UKF (Unbemanntes Kampfflugzeug: unmanned combat aircraft) has been proposed by Dornier, under the Federal German Defence Ministry's Experimental Component Programme, as a means of examining the basic problems of target acquisition, armament and flight guidance for ground support RPVs.

An Aeritalia (Fiat) G91 aircraft is being used as the testbed in this programme; in 1976 Dornier was engaged in integrating the necessary test equipment in this aircraft. VFW-Fokker is assisting in the UKF programme.

DSI/NASA OBLIQUE WING RPRA

Remotely piloted research aircraft (RPRA); first flight 6 August 1976
Power plant: One McCulloch 4318B Model O-100-1 four-cylinder piston engine (90 hp), driving a shrouded pusher propeller
Wing span: 22 ft 4 in (6.81 m)
Length overall: 19 ft 10 in (6.05 m)
Fuselage diameter: 1 ft 6 in (0.46 m)
Wing area (gross): 100.0 sq ft (9.29 m^2)
Operating weight empty: 908 lb (412 kg)
Max payload: 150 lb (68 kg)
Max T-O weight: 1,100 lb (499 kg)
Max level speed at 5,000 ft (1,525 m): 150 knots (173 mph; 278 km/h)
Max rate of climb at S/L: 1,000 ft (305 m)/min
Service ceiling: 10,000 ft (3,050 m)
Endurance: 1 hr
Launch and recovery: Conventional take-off and landing, using non-retractable tricycle landing gear. Parachute recovery system deploys automatically if command signal is lost; or can be used deliberately as alternative means of recovery.
Guidance and control: Radio/TV command guidance system. Aerodynamic control effected chiefly by elevons on the yawing wing operable differentially (as ailerons) or in unison (as elevators). Horizontal tail and vertical fin and rudder provide stability at the higher yaw angles. Aircraft designed to fly at wing yaw angles between 0° and 45°.
Special equipment: Proline command receiver; special FM data transmitter; Green Ray TV transmitter; ELT; radar transponder; flight instrumentation camera; yaw, roll and pitch gyros; UHF tracking; and vacuum pump. 1kW alternator and battery provide power for controls, TV, radio and autopilot.
Remarks: Project sponsored by NASA's Ames Research Center, with detail design and development subcontracted to DSI (Developmental Sciences Inc) in 1972. Taxying tests with prototype took place in late Autumn 1974, followed by detail modifications prior to first flight. Aircraft dismantled in Autumn 1976, after third test flight; applications of the oblique wing concept to full-size transport aircraft under consideration by NASA in 1977.

Illustrations on two following pages show the Oblique Wing RPRA with its wing in the conventional position and at maximum yaw angle of 45°

DSI RPA-9 SCOUT

Experimental mini-RPV; first flight 1975(?)
Power plant: One K & B 60 piston engine (2 hp)
Wing span: 8 ft 10.92 in (2.715 m)
Length overall: 3 ft 0 in (0.91 m)
Wing area: 14.86 sq ft (1.38 m^2)
Weight empty, equipped: 34 lb (15.4 kg)
Max launching weight: 55 lb (25 kg)
Max level speed: 60 knots (69 mph; 111 km/h)
Stalling speed: 32.5 knots (37 mph; 60 km/h)
Max rate of climb at S/L: 300 ft (91 m)/min
Launch and recovery: Normally by conventional T-O and landing. using small tricycle landing gear.
Guidance and control: VHF radio command guidance system, with 15W command output. DSI-PSA-WL-01 stability augmentation autopilot. Aerodynamic control by radio-controlled servo-operated elevons on wing trailing-edges.
Special equipment: Eight telemetry channels to monitor speed, altitude, autopilot functions and on-board sensors.
Remarks: Designed, built and flight tested initially under USAF contract. Subsequently modified and being developed under company-funded programme. A derivative of the Scout is a candidate in the USAF's LCEHV programme (which see).

DSI RPA-9 Scout in its original form

DSI RPA-12 SKY EYE

Experimental mini-RPV; first flight 26 April 1973
Data apply to Sky Eye I-B
Power plant: One McCulloch MC-101A single-cylinder piston engine (12 hp)
Wing span: 11 ft 6 in (3.50 m)
Length overall: 5 ft 7 in (1.70 m)
Wing area: 32.5 sq ft (3.02 m^2)
Max payload: 45 lb (20.4 kg)
Max T-O weight: 125 lb (56.5 kg)
Max level speed: 120 knots (138 mph; 222 km/h)
Max rate of climb at S/L: 2,200 ft (670 m)/min
Service ceiling: 15,000 ft (4,570 m)
Control-limit range: 43 nm (50 miles; 80 km)
Max endurance: 9 hr at 55 knots (63 mph; 102 km/h)
Launch and recovery: *Either* by normal T-O and landing techniques, using non-retractable tricycle landing gear; *or*, without landing gear fitted, by ramp launch and parachute recovery. Performance figures above are for latter version, in which max payload is reduced to 40 lb (18 kg).
Guidance and control: VHF radio command guidance system, based on equipment with 15W command output. Aerodynamic control by radio-control servo-operated elevons on wing trailing-edges.
Special equipment: Total-loss battery system. Alternator for flights of more than 1 hr duration. Onboard TV camera. Ciné camera can be attached externally to fuselage.

Remarks: Original Sky Eye I-A was of sweptwing tail-less configuration, with annular propeller duct aft of wing. Later modified to Sky Eye I-B, with duct removed, twin fins added near wingtips, and externally-attached ciné camera. A derivative of Sky Eye has been designed and built under LMSC contract for the US Army as the XMQM-105 Aquila (which see). Later versions are Sky Eye II-E (expendable) and II-R (recoverable), the latter being 6% larger and having endplate fins at wingtips and a lower radar cross-section.

Left: *Sky Eye I-B mini-RPV.* Right: *Sky Eye II*

E-SYSTEMS AXILLARY (USA)

Experimental mini-RPV; first flight June 1974
Wing span: 8 ft 0 in (2.44 m)
Weight: less than 60 lb (27 kg)
Loiter time: up to 6 hr
Remarks: This mini-RPV, built by E-Systems' Melpar Division originally for an anti-radiation role, is a modified version of the E-45 illustrated. Five reported to have been ordered, which were evaluated by the USAF in 1976 as potential low-cost expendable harassment vehicles (LCEHVs). The Axillary has a radar homing capability which would enable it to fulfil a defence suppression role if fitted with a small explosive charge and directed at an enemy target. It has wings with neither dihedral nor anhedral, and non-swept fins and rudders with equal area above and below the tailbooms. Other aspects of the LCEHV programme under consideration include unarmed vehicles, designed to loiter for long periods over enemy defences and impose radar silence on them; and alternative sensor payloads such as forward-looking infra-red, radiation detectors and signals jammers.

An E-Systems (Melpar) mini-RPV, of unspecified type, was reportedly to be used by the USAF in 1975 for flight demonstrations as a decoy vehicle.

E-SYSTEMS E-45, E-55 and E-100X

Experimental mini-RPVs; first flight (E-45) 1973 (?)
Data apply to (A) E-45, (B) E-100X

Power plant:
A One single-cylinder two-stroke piston engine (2 hp)
B One Rosspower four-cylinder two-stroke piston engine (9 hp)

Wing span:
A 7 ft 11 in (2.41 m)
B 10 ft 9 in (3.28 m)

Length overall:
A 7 ft 9 in (2.36 m)
B 9 ft 0 in (2.74 m)

Wing area:
A 7.36 sq ft (0.68 m^2)
B 14.33 sq ft (1.33 m^2)

Typical payload:
A 15 lb (6.8 kg)
B 50-55 lb (22.7-23.6 kg)

Typical T-O weight:
A 40 lb (18 kg)
B 100 lb (45.4 kg)

Cruising speed:
A 43.5 knots (50 mph; 80 km/h)
B 75 knots (86 mph; 139 km/h)

Rate of climb at S/L:
A 750 ft (229 m)/min
B 1,000 ft (305 m)/min

Service ceiling:
A 10,000 ft (3,050 m)
B above 10,000 ft (3,050 m)

Endurance:
A 5 hr
B more than 5 hr

Launch and recovery: Vehicle-mounted launcher or catapult. Recovery by conventional landing on ventral skid, or by net.

Guidance and control: Radio or radar command guidance system. Aerodynamic control by elevator and rudders.

Special equipment: E-Systems (Melpar Division) five-axis autopilot in E-45. Flight control system of E-100X provides remote pilot with choice of manual real-time or full autopilot control; air-to-ground telemetry is available for monitoring flight progress and navigation calculations. Wide variety of payloads possible, according to mission; these include a miniature low-cost TV reconnaissance system designed for the E-100X. All onboard electronics housed in fuselage pod.

Variants:

E-45. Twelve built, primarily as electronics testbeds, by Melpar Division of E-Systems. One of four types of mini-RPV evaluated in US Army's RPAODS programme of 1973-74.

E-55. Intended successor to E-45, with uprated engine and increased payload. Under development in 1976.

E-100X. Larger development of E-45, with more power and greater payload volume. Potential applications include TV reconnaissance, photographic reconnaissance, electronic jamming, tactical strike support, communications relay, and remote communications interception.

Illustrations on two following pages show the E-Systems E-55 and E-100X

5
U S AIR FORCE

E100X

01287

High-altitude long-endurance RPV; first flight (piloted) February 1970

Power plant: One Pratt & Whitney Aircraft of Canada PT6A-34 turboprop engine (derated to 475 ehp)

Wing span: 57 ft 0 in (17.37 m)

Length overall: 29 ft 7 in (9.02 m)

Height overall: 10 ft 8 in (3.25 m) excl antenna fairing on top of fin

Max payload: 1,100 lb (499 kg)

Max T-O weight: 5,300 lb (2,404 kg)

Cruising speed (manned): 91 knots (105 mph; 170 km/h) IAS at 45,000 ft (13,715 m)

Max rate of climb at S/L: 3,000 ft (914 m)/min

Service ceiling: above 50,000 ft (15,240 m)

Max endurance: more than 24 hr

Launch and recovery: Normal T-O and landing, using non-retractable tricycle landing gear.

Guidance and control: Radio command guidance system. Aerodynamic control by conventional ailerons, elevators and rudder. Can be controlled in take-off, constant heading/constant pitch angle, constant altitude/constant heading, constant airspeed/constant heading, reciprocal heading loiter, automatic preset heading, and landing modes.

Special equipment: 44 cu ft (1.25 m^3) payload volume for electronic relay (similar to that of a communications satellite) or data-gathering equipment. Details of equipment actually installed have not been released.

Remarks: Modified extensively by Schweizer for E-Systems from airframe of the former's SGS 2-32 sailplane, to meet military requirement. Two prototypes built, both of which have also flown with a pilot seated under a "teardrop" cockpit canopy; under- and overwing speedbrakes deleted from second aircraft, which also has increased fuel tankage. Evaluated by USAF 1971-72 in Compass Dwell programme, in competition with Martin Marietta Model 845A, but neither type was selected by USAF for production. Second prototype, in unmanned configuration, made nonstop flight of more than 21 hr at Edwards AFB on 24-25 January 1972, and followed this on 23 and 27 March 1972 with six new international altitude records and ten time-to-height records at Majors Field, Greenville, Texas. Study of potential military and civil applications, especially the latter, is continuing.

Unmanned version of the E-Systems L450F, in USAF markings

EGLEN (USAF) EAGLE (USA)

RPV
Power plant: One piston engine (13 hp), driving a two-blade pusher propeller
Wing span: 16 ft 0 in (4.88 m)
Payload: 80 90 lb (36.3 40.8 kg)
Max T-O weight: 200 lb (91 kg)
Launch and recovery: Conventional take-off and landing, using non-retractable tricycle landing gear.
Special equipment: Fully-gimballed TV relay system in nose, behind transparent nose-cap.
Remarks: Designed by USAF Flight Dynamics Laboratory. Further prototype under construction by Eglen Hovercraft Inc in early 1976.

EGLEN (USAF) FALCON

Mini-RPV
Power plant: One Homelite two-stroke piston engine (approx 3 hp)
Wing span: 8 ft 0 in (2.44 m) or 10 ft 0 in (3.05 m)
Length overall: 6 ft 0 in (1.83 m)
Wing area: 12.00 or 15.00 sq ft (1.11 or 1.39 m^2)
Max payload: 25 lb (11.3 kg)
Balanced T-O weight: 34 lb (15.4 kg)
Max T-O weight: 50 lb (22.7 kg)
Max level speed: 100 knots (115 mph; 185 km/h)
Service ceiling: approx 15,000 ft (4,575 m)
Max endurance: 5 hr
Launch and recovery: Conventional T-O and landing, using non-retractable tricycle landing gear with steerable nosewheel. Wheels can be replaced by skis for snow or desert operation. Capable also of shipboard operation using a catapult launch and net recovery system.
Guidance and control: Radio command guidance system. Aerodynamic control by ailerons, elevators and rudder.
Special equipment: Six-channel Proline radio control system standard (four channels for flight control and two for payload operation). Eight-channel system available optionally. Standard airborne TV payload package includes Conic 8W CTM UHF 408V S-band TV transmitter with antenna; Sony AVC 1400 TV camera and lens (modified for airborne use), and battery pack. Each TV payload package is modified to accept the microwave transmission system utilised, and is fully systems-tested.
Remarks: Designed by USAF Flight Dynamics Laboratory. In service with several military and civil authorities, including NATO forces, for testing sensor and weapon systems and for weather monitoring. Available also in target version. Other potential applications include reconnaissance, surveillance, photography, search missions, flood or forestry survey, air sampling, traffic observation, infra-red or laser target designation, and air testing of autopilots, flight computers and other equipment. Construction, mainly of glassfibre, foam plastics and plywood, with epoxy paint finish, results in minimal radar signature.

Left: *First prototype Eagle, built by the USAF.* Right: *Falcon multi-purpose mini-RPV*

FAIRCHILD SAILWING

Experimental RPV
Power plant: One McCulloch MC-101A single-cylinder piston engine (12 hp)
Wing span: 14 ft 8½ in (4.48 m)
Length overall: 11 ft 10 in (3.63 m) excl nose probe
Wing area: 33.7 sq ft (3.13 m^2)
Payload: 50 lb (22.7 kg)
Max T-O weight: 250 lb (113 kg)
Performance: No details known
Launch and recovery: Normal T-O and landing techniques, using non-retractable tricycle landing gear.
Guidance and control: Radio command guidance system. Aerodynamic control by ailerons (on some models), elevators and rudder.
Special equipment: Command, autopilot and telemetry systems standard; 28V 840W alternator fitted in some models. RPAODS test version carried a three-axis autopilot and Vega C-band command and control system. Details of mission equipment not disclosed, but payloads flown have included photographic, electro-optical, infrared, early-warning and communications relay systems.

Remarks: Basic design utilises a specially-shaped fabric wing glove over a rigid leading-edge and tip rib, and is cable-tensioned along the trailing-edge. Wings are wire-braced, and can be folded back alongside fuselage to facilitate transport, handling and storage. Various Sailwing models were built, and were flown in such defence programmes as Black Fly and Dragon Fly (USAF), Lookout (US Navy) and RPAODS (US Army). All Sailwing programmes are now inactive. A much smaller variant, named **Minipop**, was built for the US Navy but did not go into production.

Left: *Sailwing RPV with tricycle landing gear.* Right: *With lightweight jettisonable landing gear. Latter view also shows wings folded*

A further variant of the Sailwing RPV on the pneumatic launcher developed under the US Navy's Project Lookout

Sailwing recovery system, using a vertical barrier net to arrest the RPV and a horizontal catch net to support it

FLIGHT REFUELLING FR-500

Multi-purpose target drone
Power plant: One Noel Penny Turbines NPT 401 turbojet engine (300 lb; 136 kg st); three standard JATO bottles (each 2,645 lb; 1,200 kg st) for launch boost, with 0.7 sec burn
Wing span: 7 ft 0 in (2.13 m)
Length overall: 14 ft 0 in (4.27 m)
Height overall: 3 ft 6 in (1.07 m)
Basic operating weight empty: 265 lb (120 kg)
Payload volume: 0.75 cu ft (0.02 m^3)
Payload with full fuel load: 35 lb (16 kg)
Max launching weight: 490 lb (222 kg)
Max level speed at S/L: 500 knots (575 mph; 926 km/h)
Max rate of climb at S/L: 8,800 ft (2,682 m)/min
Time to 25,000 ft (7,620 m): 5 min
Service ceiling: 40,000 ft (12,200 m)
Max range: more than 600 nm (691 miles; 1,112 km)
Endurance at S/L:
at 300 knots (345 mph; 556 km/h) more than 1 hr
at 500 knots (575 mph; 926 km/h) 30 min
Launch and recovery: Launched by JATO booster rockets. Automatic or command parachute recovery system. Built-in buoyancy to permit recovery from water.
Guidance and control: Radio command guidance system. Aerodynamic control by ailerons, elevators and rudder.
Special equipment: Basic control and tracking equipment similar to that installed in the Northrop MQM-74 targets currently in use by Royal Navy, using Marconi-Elliott autopilot and VHF receiver/decoder. Provision to replace this with smaller and lighter Vega 656 system, which would also increase available payload volume and weight to 1.5 cu ft (0.04 m^3) and 40 lb (18 kg) respectively. Other equipment includes 28V electrical system (engine-driven alternator and nickel-cadmium battery) and recovery beacon. Operational equipment, according to mission, can include up to six infra-red/tracking flares on underwing pylons; nose and/or tail mounted Luneberg lenses for radar augmentation; miss-distance and scoring equipment; radio altimeter; underwing towed targets, cable reel and brake; and reconnaissance/surveillance sensors.
Remarks: Being developed under Ministry of Defence contract. Designed primarily for use by maritime forces for gunnery and weapons training at sea, but capable of adaptation to various other roles.

Artist's impression of the Flight Refuelling FR-500 target

PHILCO

FORD CALERE and PRAEIRE

Mini-RPVs for battlefield reconnaissance, target acquisition and laser designation

Data apply to Praeire IIB

Power plant: One Kolbo Korp two-cylinder piston engine (8 hp)

Wing span: 13 ft 0 in (3.96 m)

Length overall: 11 ft 0 in (3.35 m)

Payload (TV sensor system): 25 lb (11.3 kg)

Max T-O weight: 135 lb (61.2 kg)

Max level speed: 74 knots (85 mph; 137 km/h) at 5,000 ft (1,525 m)

Ceiling: 10,000 ft (3,050 m)

Mission radius: Radio line-of-sight

Mission endurance: 4 hr

Launch and recovery: Conventional T-O and landing, using non-retractable tricycle landing gear; or catapult launch.

Guidance and control: Radio command guidance system. An Earth's electric field autopilot provides attitude control when out of sight of controller.

Special equipment: Ford stabilised TV sensor package contains high-resolution camera with 4:1 zoom capability sightline. Coelostat mirror optical system provides 360° azimuth and +5°/−75° elevation coverage for pointing and stabilisation. International Laser Systems miniature laser provides target ranging and designation through a common optical path with the TV. Sensor system contained in nose module which rotates 180° after take-off to locate optical dome underneath for viewing. Electronic warfare payload for Praeire has been developed by Ford, for tactical communications interception and jamming. Calere versions incorporate FLIR (forward-looking infra-red) for full night-time capability. Ford has developed a miniature FLIR sensor which weighs only 6 lb (2.7 kg) and has three fields of view.

Variants:

Praeire I. Original version, developed for USAF in 1973. Two-cylinder nose-mounted engine, 10 ft (3.05 m) wing span, 25 lb (11.3 kg) payload comprising TV camera and laser transmitter. Of simple high-wing layout, with fixed tailwheel landing gear.

Praeire II. Larger vehicle, of completely new design, with mid-mounted engine and pusher propeller. Evaluated in US Army RPAODS programme in 1974.

Praeire IIB. Improved version of Praeire II, as described. Glassfibre/polyurethane foam construction. Under evaluation by US Army in 1976, including successful designation of a tank target for the Cannon-Launched Guided Projectile. Some have been ordered by Israel.

Calere. Night-operating equivalents of Praeire models, to which they are generally similar except for FLIR installation.

Left: *Current version of Praeire IIB, with Kolbo Korp engine*

Surveillance mini-RPV; first flight 1976
Power plant: One Kolbo Korp two-cylinder piston engine (5 hp)
Wing span: 9 ft 0 in (2.74 m)
Payload: 20 lb (9.1 kg)
Max T-O weight: 60 lb (27.2 kg)
Performance: No details known
Special equipment: Payload includes lightweight FLIR (forward-looking infra-red) package, optical surveillance devices and International Laser Systems laser target designator.
Remarks: Being developed for US Army's LANDSS (Lightweight Advanced Night/Day Surveillance System) programme under an ARPA (Advanced Projects Research Agency) contract. Two prototypes ordered, of which flight testing was completed in 1976.

This Ford Praeire IIB demonstrated successful anti-tank operation in conjunction with the US Army's Cannon Launched Guided Projectile (CLGP). The Calere III will be a smaller night-flying variant of the same basic design

FRANCE-ENGINS MITSOUBAC

Decoy or target drone; first flight 1975
Power plant: One Microturbo TRS 18-056 turbojet engine (220 lb;100 kg st)
Wing span: 4 ft 5 in (1.35 m)
Length overall: 9 ft 4½ in (2.86 m)
Height overall: 2 ft 4½ in (0.72 m)
Weight empty: 183 lb (83 kg)
Max launching weight:
expendable target 220.5 lb (100 kg)
recoverable target 264.5 lb (120 kg)
decoy 297.5 lb (135 kg)
Payload (all versions):
with parachute recovery system 55 lb (25 kg)
without parachute recovery system 77 lb (35 kg)
Max speed:
expendable target Mach 0.85 at 30,000 ft (9,145 m)
recoverable target Mach 0.6-0.75 between 150-900 ft (45-275 m)
decoy Mach 0.7 at low altitude
Endurance at max speed:
expendable target 15 min
Range:
recoverable target 108-162 nm (124-186 miles; 200-300 km)
decoy 162 nm (186 miles; 300 km)
Launch and recovery: Design permits parachute recovery, catapult launch, or air launch from a helicopter or a fixed-wing aircraft; in the latter case the Mitsoubac is carried inverted on the launch aircraft's external stores pylons.
Guidance and control: Can be pre-programmed, or controlled in flight via an onboard light aircraft autopilot with a vertical gyro. Elevons actuated by pneumatic jacks.
Remarks: Envisaged in several versions, including those of expendable or recoverable target drone and defence saturation decoy. Airframe is of moulded laminated polyester. Flight testing was continuing in 1976.

Illustrations on two following pages show a model of Mitsoubac and a full-scale prototype launched from a Bell 47 helicopter

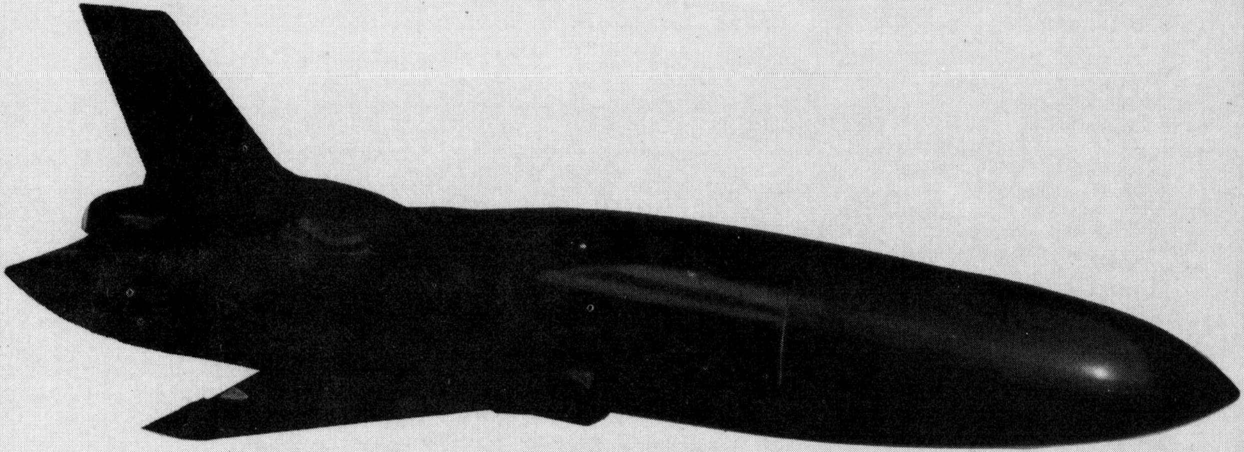

A92
255

Target drone; first flight (Mk 1) 28 August 1952
Data apply to Mk 3B standard-span version with Mk 8 wingtip pods
Power plant: One Rolls-Royce Viper Mk 201 turbojet engine (2,500 lb; 1,134 kg st)
Wing span: 20 ft 8.99 in (6.32 m)
Length overall (excl nose probe): 23 ft 3 in (7.11 m)
Wing area: 76.00 sq ft (7.06 m^2)
Min weight empty, equipped: 2,900 lb (1,315 kg)
Max payload: 550 lb (249 kg)
Max T-O weight: 3,650 lb (1,655 kg)
Max level speed: 490 knots (564 mph; 908 km/h) at 54,000 ft (16,460 m)
Time to 54,000 ft (16,460 m): 30 min
Typical max on-station endurance: 1 hr 38 min
Max range: 670 nm (771 miles; 1,240 km)
Launch and recovery: Mounted for take-off on tubular-framed tricycle trolley which accelerates under drone's normal jet power and is jettisoned when drone rotates. Landing on retractable underfuselage skid and steel auxiliary wingtip skids.
Guidance and control: Radio command guidance system, comprising two receiver/selectors (second of which may be used for standby or destruct) and GAF relay receiving set. Telemetry equipment consists of NIC transducers and Australian-designed transmitter and junction box. Aerodynamic control by flaps, ailerons and elevators; underfuselage airbrake optional.
Special equipment: Transponders and microwave reflectors for trials with active, semi-active or beam-riding missiles. Heat sources, including infra-red flare packs in rear of fuselage, can be fitted for low-frequency IR output. X-, S- and C-band transponders can be fitted for target acquisition and to enable Jindivik to be tracked at greater range. Provision for "Tonic" winch pylon under each wing, on which can be carried a recoverable active radar or infra-red "Tonic" CG tow "bird" or expendable tow bird with infra-red-augmented nose. These can be towed at 50-500 ft (15-150 m) behind aircraft; recovery is by electric winch. Other types of recoverable towed target can carry augmentation in form of transponders, microwave reflectors or infra-red. Wide-angle cameras in wingtip pods, with all-round vision, film and record approach path and proximity of missiles fired against target. Pod variants are Mk 5 (cameras only), and Mk 8 (cameras, fuel and provision for microwave reflectors in leading-edge and trailing-edge radomes). Rearward-facing cameras, and a Luneberg lens reflector mounted in fairing above jetpipe, may be used to film and record operations with towed targets. Both Mk 5 and Mk 8 pods usable on standard wings for low/medium altitude work, or on 26 ft 6 in (7.92 m) extended wings for high altitudes. Further wing extension to 32 ft 1.4 in (9.78 m) available for extra high altitude work, with which Mk 5 pods only are used; this version also fitted with ventral tail-fin.

Jindivik Mk 3B with high-altitude wing extensions

Variants:
Pika. Prototype, to MoS Specification E.7/48. Armstrong Siddeley Adder ASA.1 turbojet (1,050 lb; 476 kg st) and provision for pilot. Span 19 ft 0 in (5.79 m), length 23 ft 3 in (7.11 m).
Mk 1. Adder engine. First flown 28 August 1952: 14 built.
Mk 2. Viper ASV.3 engine (1,900 lb; 862 kg st). First flown 11 December 1953: 111 built.
Mk 2A. Modified wings and intake. First flown 18 September 1958: 3 built.
Mk 2B. Viper ASV.8 engine. First flown 8 October 1959: 76 built.

Mk 3. Viper ASV.11 engine. First flown 12 May 1961: 9 built.
Mk 3A. First flown 10 November 1961: 147 built.
Mk 3B. Current production version, as described. First flown 22 January 1970: 126 ordered by January 1977, of which 79 delivered. Built for UK (90 Mk 103B), Royal Australian Navy (27 Mk 203B) and Weapons Research Establishment, Woomera.
Remarks: Total of 486 Jindiviks ordered by January 1977, including 234 for UK, 163 for joint UK/Australian weapons project, 42 for US Navy, 37 for Royal Australian Navy and 10 for Sweden. Built by Australian Government Aircraft Factories, but those used in UK are assembled and equipped by BAC Guided Weapons Division. Jindivik is also being used as basis of a Bell Aerospace/USAF study for an air cushion landing system (ACLS) for RPV applications; a GAF-modified Jindivik Mk 203A, fitted with an ACLS, completed taxying trials in May 1976 at speeds of up to 50 knots (57.5 mph; 92.5 km/h). Flight trials expected to begin in 1977.

Left: *Royal Australian Navy Jindivik Mk 203A fitted with air cushion landing system*

Above: *Jindivik Mk 3B with wings of standard span fitted with wingtip pods*

TURANA
GAF

GAF TURANA

Target drone; first flight 12 March 1971
Power plant: One Microturbo Couguar 022 turbojet engine (176 lb; 80 kg st); one PMD41 single-nozzle solid-propellant booster motor (6,000 lb; 2,721 kg nominal thrust, 2 sec nominal burning time)
Wing span: 5 ft 0.2 in (1.53 m)
Length overall: 11 ft 0½ in (3.37 m)
Height: 3 ft 10.8 in (1.19 m) incl booster; 3 ft 4 in (1.02 m) excl booster
Wing area: 13.2 sq ft (1.23 m^2)
Weight dry (excl special equipment): 364 lb (165 kg)
Weight at launch (excl special equipment): 615 lb (279 kg)
Weight of special equipment: more than 100 lb (45 kg)
Max level speed: 390 knots (450 mph; 724 km/h)
Nominal boost acceleration: 10*g*
End-of-boost speed: 560 ft (170 m)/sec
Service ceiling: 33,000 ft (10,000 m)
Max rate of climb at S/L: 4,000 ft (1,220 m)/min
Range: 325 nm (374 miles; 602 km)
Endurance: 55 min
Launch and recovery: Launched from standard Ikara ship launcher, or from lightweight portable launcher, at fixed elevation of 22° 30′. Adaptable for air launch. Command parachute recovery, activated automatically in the event of engine, electrical power system or command link failure.
Guidance and control: Initial guidance equipment adapted from Ikara system on board RAN ships. Navigation by Ikara tracking receiver and ship's plotting facilities. Control by open loop demands from ground and closed loop sensing and autocontrol system in the drone. Steering by autopilot and electrically-actuated elevons. Air data unit, height lock and speed lock loops standard; radar altimeter, for low-altitude operation, optional.
Special equipment: Smoke release system for visual augmentation; 7½ in (190 mm) Luneberg lens in nose for forward-looking passive radar augmentation. Other equipment may include a 465MHz 48-channel telemetry system, pyrotechnic flares, acoustic miss-distance indication system, or a short-tow system to carry infra-red flares or a Luneberg lens radar reflector.
Remarks: Development of Ikara anti-submarine missile for Royal Australian Navy, which has ordered 12. Deliveries began in late 1973. Service trials continuing in 1977; expected to become operational by end of 1977.

Turana with underfuselage booster in place

HAWKER DE HAVILLAND ENMOTH

Target system; first flight 1975
Data apply to HDH-10 target aircraft
Power plant: One K & B 61 piston engine (2 hp)
Wing span: 5 ft 0 in (1.52 m)
Length of fuselage: 5 ft 2 in (1.57 m)
Height overall: 1 ft 10 in (0.56 m)
Weight empty: 15 lb (6.8 kg)
Max normal payload: 6.75 lb (3.0 kg)
Max T-O weight: 25 lb (11.3 kg)
Max speed: 70.5 knots (81 mph; 130.5 km/h)
Stalling speed: 20 knots (23 mph; 37 km/h)
Max endurance: 52 min at 50% power
Launch and recovery: Conventional take-off and landing, using non-retractable tricycle landing gear with steerable nosewheel. Parachute recovery available optionally.
Guidance and control: Radio command guidance system. Aerodynamic control in pitch and roll by elevons on wing trailing-edge. Rudder is fitted primarily to facilitate crosswind take-offs and landings. Aircraft is virtually unspinnable. Futaba flight controls and actuators.
Special equipment: Infra-red source mounted on top of fin. Normal configuration, at minimum speed of 20 knots, is for a combined payload/fuel capability of 10 lb (4.5 kg), including a maximum 3.25 lb (1.5 kg) of fuel; combined capacity can be increased to over 11 lb (5 kg) at min speed of 30 knots (34.5 mph; 55.5 km/h). Airframe (production version) of glassfibre-covered polystyrene foam.
Variants:
HDH-10. Miniature target drone, as described. Complete Enmoth system, developed for Royal Australian Army to provide aiming and scoring system for Redeye surface-to-air missiles, comprises HDH-10 target aircraft, transport container and self-contained ground support package. Twelve ordered in January 1977.
HDH-11 Beemoth. Projected development of HDH-10, having a similar planform, with double the linear dimensions of Enmoth; a pusher propeller; and greatly increased payload.

HDH-10 Enmoth

KAMAN STAPL (USA)

Experimental surveillance/target acquisition RPV; first flight scheduled for mid-1977.

Under contract to the US Office of Naval Research, Kaman has designed, built and ground tested two prototypes of the STAPL (Ship Tethered Aerial Platform); preliminary design was carried out in 1970 under a previous ONR contract. Testing of the two aircraft, their mobile launch and recovery platform, and the control system, began in 1974 and was to continue during 1977. Modifications are incorporated in the second prototype.

The prototype aircraft are of autogyro configuration, and are equipped with automatic flight control systems and data recording equipment. The self-contained AFCS, engineered by Kaman, provides three-axis stabilisation and automatic flight path control, and incorporates redundancy for mission reliability.

Kaman is also conducting a number of independent design studies and mission analyses for rotating-wing RPVs. One of these, designated K-244, competed in 1974 in the now-abandoned US Army RPAODS programme. An aircraft similar in concept to STAPL is under study for a battlefield surveillance/target acquisition role.

First prototype of the Ship Tethered Aerial Platform (STAPL)

LAVOCHKIN La-17

This small radio-controlled target drone is used in the USSR for training surface-to-air missile crews and the pilots of interceptor fighter aircraft. It has a slim cylindrical fuselage, with unswept wings of constant chord mid-set on the fuselage at about mid-length. The conventional tail surfaces are also unswept and of constant chord, with the tailplane mounted partway up the fin to avoid the exhaust from two jettisonable solid-propellant booster rockets which are fitted under the wings for launch. The La-17, which was developed in the early 1950s, has an all-metal airframe. Cruise propulsion is provided by a turbojet engine mounted in a pod under the fuselage.

Launch photograph of an La-17 with the boosters firing

LMSC AEQUARE (USA)

Expendable reconnaissance and laser target designator mini-RPV; first flight mid-1975

Power plant: One McCulloch MC-101 single-cylinder piston engine (12.5 hp), driving a three-blade ducted pusher propeller

Wing span: approx 12 ft 0 in (3.66 m)

Length overall: approx 8 ft 0 in (2.44 m)

Max launching weight: 150 lb (68 kg)

Range: nearly 174 nm (200 miles; 322 km)

Launch and recovery: Air-launched at 25,000 ft (7,620 m) from fighter aircraft (F-4 Phantom in tests), under starboard wing of which it is carried with wings folded inside a modified SUU-42 flare dispenser pod. A CTV-2 pod containing the command/telemetry/video data link is carried under the fighter's port wing. Upon launch, end cap of carrier pod is released by explosive charge, deploying a drogue parachute on which pod descends to 14,000 ft (4,270 m). Barometric device then releases main chute, which pulls pod clear of RPV. Wings of Aequare then pivot forward into flight position, and engine is started. Provision also made for ground launch by rocket. RPV airframe is expendable, but parachute system is provided for recovery of payload package.

Guidance and control: After launch of pod, carrier aircraft climbs to 35,000 ft (10,670 m) and acts as relay between RPV and ground control station. On release from pod, RPV is pre-programmed to enter level, powered flight and to set up data link with ground controller. Remainder of flight is normally controlled from ground station, but can be pre-programmed if desired.

Special equipment: Recoverable payload compartment in lower fuselage contains Ford stabilised TV (non-stabilised in initial tests) and laser target designator, Aacom data link equipment, Lockheed autopilot, and X- and C-band beacons.

Remarks: "Feasibility" RPV for USAF evaluation. Tests scheduled for 1976 included use in connection with launch of 500 lb Mk 82 laser-guided bombs. Glassfibre airframe built by Windecker Industries Inc.

LMSC XMQM-105 AQUILA

Technology demonstrator for battlefield surveillance/target designation mini-RPV; first flight December 1975
Power plant: One McCulloch MC-101M/C single-cylinder piston engine (11 hp)
Wing span: 11 ft 10.8 in (3.63 m)
Length overall: 6 ft 0 in (1.83 m)
Payload: 30 lb (13.6 kg)
Max launching weight: 120 lb (54.4 kg)
Max speed: 118 knots (136 mph; 219 km/h)
Service ceiling: above 20,000 ft (6,100 m)
Max endurance: 3 hr
Launch and recovery: Conventional T-O and landing during initial flight tests, using non-retractable tricycle landing gear. Operationally, is launched by truck-mounted pneumatic-tube launcher at airspeed of 44 knots (51 mph; 81.5 km/h). Initially, was recovered by engaging extendable arrester hook in parallel-strap retrieval net; this system superseded in 1976 by direct flight into recovery net.
Guidance and control: Radio/TV command guidance system. Automatic terminal control system. Aerodynamic control by elevons on wing trailing-edges.
Special equipment: Lockheed autopilot, Aacom data link, Honeywell laser designator and Perkin-Elmer reconnaissance camera. Five different payloads to be evaluated: for real-time surveillance, photographic reconnaissance (Perkin-Elmer 35 mm still camera), target acquisition (TV camera), laser target designation (Honeywell designator), and target location and artillery adjustment. Crash-resistant payload protection system standard.
Remarks: Derived from DSI Sky Eye (which see), and developed by DSI for the US Army under subcontract to LMSC. Half-scale test model flown for first time on 14 July 1975. Thirty Aquilas, and four ground stations, being built for systems technology demonstration, of which 13 Aquilas delivered by early 1976.

Photographs overleaf

Left: *LMSC XMQM-105 Aquila on pneumatic rail launcher* Above: *Aquila mini-RPV photographed after recovery*

MARCHETTI HELISCOPE

Electrically-powered flying platform

Marchetti has undertaken design studies of a variety of rotating-wing aircraft in recent years, under contract from the French Direction des Recherches et Moyens d'Essais (DRME). These were continuing in 1976.

The Heliscope flying platform consists basically of two three-blade light alloy fixed-pitch rotors of co-axial contra-rotating design, attached respectively to the rotor and stator of an electric motor located between their hubs. The power supply for this motor is provided by a generator mounted on the vehicle which serves as the mobile ground station, and is fed through a cable which runs to the centre of the base of the Heliscope.

Marchetti has used test results to project a platform using three Heliscopes, each with a four-blade rotor of 5 ft 3 in (1.60 m) diameter. Such a vehicle could support a useful load of 220 lb (100 kg) with a simplified stabilisation system at an altitude of 755 ft (230 m) above the ground. Heliscopes with higher performance, and new electrical free-flying platforms, are under study.

Left: *Motorised mockup of a flying platform with three Marchetti Heliscopes.* Above: *Heliscope in simulated hovering flight, suspended from a crane*

MBLE ÉPERVIER (SPARROWHAWK)

Battlefield surveillance system; first flight (X-1) 24 April 1965
Data apply to Épervier RPV
Power plant: One Lucas CT3201 turbojet engine (114 lb; 51.7 kg st)
Wing span: 5 ft 7 in (1.72 m)
Length overall: 7 ft 9 in (2.38 m)
Height overall: 3 ft 0 in (0.92 m)
Payload: approx 45 lb (20 kg)
Max launching weight: 313 lb (142 kg)
Cruising speed: 270 knots (310 mph; 500 km/h)
Max mission radius: 40 nm (47 miles; 75 km)
Endurance: more than 25 min
Launch and recovery: From short orientatable ramp, with associated checkout device. Parachute recovery system, with guaranteed CEP of 500 ft (150 m).
Special equipment: Omera 5 in, or Omera or Oude-Delft 70 mm, day or night cameras; photoflash gear for night operation; SAT Cyclope infra-red line-scanning equipment, or Teledyne Brown low light level TV, both with real-time transmission, can be carried. Capable of photographing pinpoint targets or large areas of terrain during guided and/or programmed flights up to more than 38 nm (43.5 miles; 70 km) from base.
Remarks: Épervier is the name for the complete surveillance system which includes the drone vehicle, its launcher, and a drone control centre (DCC). The DCC contains all the necessary electronic equipment for guiding and tracking. A mobile unit for photographic processing and interpretation also forms part of the system. Initially developed to meet NATO specifications, the Épervier system received Belgian government financial support in mid-1969, and in 1974 an initial order was placed on behalf of the Belgian Army for the production, in two batches, of 43 Éperviers and associated ground equipment. Two complete squadrons were delivered in 1976, at which time MBLE had announced a simplified version, for export to non-NATO countries, known as the **Asmodée**.

Epervier battlefield surveillance RPV

(Italy)

METEOR GUFONE (OWL)

Tactical reconnaissance RPV; first flight 1972

Power plant: One Williams Research WR24-6 turbojet engine (121 lb; 55 kg st), and two Meteor 8785/CNS jettisonable solid-fuel booster rockets (total thrust 5,730 lb; 2,600 kg for 0.7 sec)

Wing span: 5 ft 6 in (1.69 m)

Length overall: 11 ft 10 in (3.61 m)

Height overall: 2 ft 5 in (0.73 m)

Typical payload: 66 lb (30 kg)

Max launching weight: 300 lb (136 kg)

Max level speed: 400 knots (460 mph; 740 km/h)

Max cruising height: 35,000 ft (10,670 m)

Range penetration:

at S/L 29-48 nm (34-56 miles; 55-90 km)

at 32,800 ft (10,000 m) 72-113 nm (83-130 miles; 135-210 km)

Launch and recovery: Launch by booster rockets from Jeep-mounted launcher. Parachute recovery.

Guidance and control: Radio command guidance system. Flight control by conventional ailerons and elevators. Once vehicle is on course, it becomes "deaf" to all friendly or enemy electronic signals until nearing the end of its return flight, when it again comes under command guidance for recovery. Provision for intermediate pre-programmed guidance phase between guided and "deaf" phases, in which event vehicle will accept only specially coded commands of very short duration, for sole purpose of course correction.

Special equipment: Equipment for day and night operations can include a variety of infra-red sensors and cameras using 50, 70 or 75 mm film to photograph a strip of terrain more than 55 nm (63 miles; 102 km) long and (according to version) 3,280 ft (1,000 m), 6,560 ft (2,000 m) or 9,840 ft (3,000 m) wide, respectively from altitudes of 1,000 ft (305 m), 2,000 ft (610 m) or 3,000 ft (915 m). For night operations, can carry 14 wingtip flares which are dropped at preselected intervals of time.

Remarks: Gufone is the operational vehicle element of the Gufo system which also comprises sections for launching; guidance and control; sensor recovery, interpretation and headquarters; and vehicle recovery and preparation. The Gufone RPV is based upon the Northrop Chukar I (MQM-74A, which see), modified by Meteor. The Gufo system has been evaluated successfully; approval for production for the Italian armed forces is awaited.

Gufo system control truck and Gufone drone on towed launcher

Multi-role RPV; first flight 1975 (?)
Data apply to version with 330 lb (150 kg) st engine
Power plant: One turbojet engine of unspecified type (330 lb; 150 kg st); provision for alternative turbojet of 660 lb (300 kg) st
Wing span: 7 ft 10½ in (2.40 m)
Length overall: 13 ft 9 in (4.20 m)
Weight empty: 401 lb (182 kg)
Max internal payload: 128 lb (58 kg)
Max launching weight without external loads: 758.5 lb (344 kg)
Max weight of external loads: 220.5 lb (100 kg)
Max speed: 442 knots (509 mph; 820 km/h) at 29,525 ft (9,000 m)
Service ceiling: 39,375 ft (12,000 m)
Out-and-back range at 29,525 ft (9,000 m): 256 nm (295 miles; 475 km)
Endurance:
at max speed at S/L 30 min
at econ cruising speed of 372 knots (429 mph; 690 km/h) at 29,525 ft (9,000 m) 1 hr 24 min
Launch and recovery: Ground launch from zero-length launcher Self-recoverable.
Special equipment: **Alamak** electronic ground and airborne subsystem for guidance, tracking, plotting, telemetry and/or navigation on training missions. More advanced subsystem, known as **Sirak**, for combat missions in high-jamming areas.
Remarks: Mirach is the airborne vehicle component of an overall multi-role RPV system named **Andromeda**, which comprises also the Alamak or Sirak subsystems described in the previous paragraph, and a ground support system suitable for training or combat missions. The Alamak and ground support are already in production and service with other systems; the Sirak subsystem was in an advanced development phase in 1976. The Mirach is of modular construction, to facilitate the interchangeability of parts, and can be fitted with a 661 lb (300 kg) st turbojet engine giving it a maximum speed of about 540 knots (621 mph; 1,000 km/h).

Mirach RPV component of Meteor Andromeda *system*

NAVY

METEOR P.1/P.2 series and P.X

Target or reconnaissance drone; first flight 1959 (?)
Data: P.1 out-of-sight version, 120 hp engine
Power plant: One Meteor Alfa 1AQ four-cylinder piston engine (120 hp)
Wing span (without wingtip containers): 12 ft 1 in (3.68 m)
Length overall: 11 ft 1½ in (3.39 m)
Height overall: 2 ft 1½ in (0.65 m)
Weight without fuel and electronics: 293 lb (133 kg)
Launching weight:
1 hr flight 444 lb (202 kg)
max 495 lb (225 kg)
Max level speed at 21,325 ft (6,500 m): 296 knots (342 mph; 550 km/h)
Max rate of climb at S/L: 3,940 ft (1,200 m)/min
Time to 20,000 ft (6,100 m): 5 min
Max range: 86 nm (100 miles; 160 km)
Launch and recovery: Ramp, catapult or air launch. Parachute recovery, from land or water.
Guidance and control: UHF radio command guidance system. Onboard receiver translates signals into operations controlling elevators and ailerons, stopping engine, and deploying recovery parachute. Ailerons and elevators operated by electrical servo controls.
Special equipment: Two-axis automatic stabilisation system, plus gyro for lateral stabilisation. Wingtip reflectors can be fitted to amplify echoes from target. Alternative UHF tracking and tele-control system can be used, in conjunction with a 4.4 lb (2 kg) transponder in the target vehicle.
Variants:
P.1. Built in two versions, with 100 hp Meteor Alfa 1 engine or (as described) 120 hp Alfa 1AQ; either version can be used for in-sight and out-of-sight operation. Production completed.
P.1/R. Reconnaissance version of P.1, with 110 hp Meteor Alfa 1 engine and 3,968 lb (1,800 kg) st Meteor 8785/Z solid-fuel booster rocket. Built for Italian armed forces. Production completed.
P.2. Slightly larger than P.1, primarily for out-of-sight operation. Built in two versions, with 160 hp Alfa 3AQ or 320 hp Alfa 5 engine. Production completed.
P.X. Generally similar to P.1, with 72 hp McCulloch O-100-1 engine and Meteor 8785 booster. Designed for zero-length launch and radar tracking. Built for Italian and other armed forces; in production.

Naval version of the Meteor P.1 target on zero-length launcher (far left) and the McCulloch-engined P.X (left)

NASA MINI-SNIFFER I and II

Remotely piloted research vehicle; first flight August 1974
Data: Mini-Sniffer II (mission vehicle); performance data estimated
Power plant: One Akkerman 235 piston engine (15 hp), using hydrazine monopropellant
Wing span: 18 ft 0 in (5.49 m)
Max payload: 50 lb (22.7 kg)
Max T-O weight: 145 lb (65.8 kg)
Max level speed: 160 knots (184 mph; 296 km/h) at 70,000 ft (21,340 m)
Max rate of climb at S/L: 2,600 ft (792 m)/min
Service ceiling: 90,000 ft (27,430 m)
Range with max payload: 434 nm (500 miles; 805 km)
Launch and recovery: Conventional T-O and landing, using non-retractable tricycle landing gear, without wheel tyres.
Guidance and control: Radio command guidance system. Aerodynamic control by ailerons, rudders and elevator.
Special equipment: Alternator on engine for electrical power during climb and cruise; 28V nickel-cadmium battery for use during unpowered flight; Resdel uplink control; SCI-680-PCM 16-channel data telemetry system; Vega radar transponder; NASA autopilot; Sun and magnetic sensors for wing levelling, airspeed hold and heading hold. Payload instrumentation to measure nitric oxide, chlorine, Freon 11 and 12, free oxygen, and the hydroxyl radical. Lightweight wire impact particle grab samplers, and real-time telemetered particulate density sensor. Radiation sensor payloads under development.
Variants:
Mini-Sniffer I. First prototype, powered by a 7 hp McCulloch 91 piston engine (fuel/oil mixture). Conventional T-O and landing (non-retractable tricycle gear, without wheel tyres). Used to flight-test airframe for aerodynamic, guidance and control data, and to develop launch and recovery techniques. Sweptwing configuration, with endplate fins and rudders and canard foreplanes.
Mini-Sniffer II. Mission vehicle, first flown in July 1975 with McCulloch engine. Twin tailbooms, twin fins and rudders, enclosed tailplane and variable-dihedral wingtips. Wingtip fins and rudders and canard foreplanes deleted. Otherwise similar in overall appearance to Mini-Sniffer I. Fitted in 1976 with Akkerman hydrazine engine and mission payloads, to carry out high-altitude atmospheric research (up to 100,000 ft; 30,500 m) in remote areas.

Left: *Mini-Sniffer II, with variable-dihedral wingtips.* Right: *Mini-Sniffer II, with Mini Sniffer I prototype in background*

"REPEATER"

(USA) NORTHROP FALCONER/MQM-57A

Surveillance drone; first flight 1955
Power plant: One McCulloch O-100 four-cylinder piston engine (92 hp)
Wing span: 11 ft 6 in (3.50 m)
Length overall: 13 ft 5 in (4.09 m)
Height overall: 2 ft 7½ in (0.80 m)
Weight empty: 354 lb (161 kg)
Max payload: 75 lb (34 kg)
Max launching weight: 442 lb (200 kg)
Max level speed: 174 knots (201 mph; 323 km/h) at S/L
Service ceiling: 15,000 ft (4,570 m)
Endurance at S/L: 35 min
Launch and recovery: Launched from zero-length launcher by two M-3 JATO bottles mounted on fuselage sides. Recovery by single-stage 44 ft (13.4 m) diameter extended-skirt parachute, deployed on command, automatically on loss of radio command, or if parachute safety channel ceases to function. Integral keel absorbs landing impact.
Guidance and control: Radio command guidance system, incorporating AN/ARW-79 autopilot. Radar tracking. Aerodynamic control by ailerons, elevators and rudder.
Special equipment: Still (two 75 mm) or TV surveillance cameras for day and night photographic intelligence and reconnaissance; flares for night photography. Radar beacon or passive radar reflector pods and night light identification system. Cameras can be replaced by other sensor systems. All operational equipment loaded through hatches in top of fuselage.
Variants:
Falconer. Developed for US Army Signal Corps from KD2R-5 (see entry for Northrop MQM-36). Known originally as AN/USD-1, later as MQM-57A. Built for US Army and several foreign countries.
NV-104. Experimental, much-modified and refined version of MQM-57A, with 250 shp Allison T63-A-5 turboshaft engine and 150 lb (68 kg) payload capacity. Evaluated 1964 but not built in quantity.

Falconer surveillance drone on its lightweight launcher

NAVY

NORTHROP MQM-36 and KD2R-5

Training target drone; first flight 1947
Data apply to KD2R-5
Power plant: One McCulloch O-100-3 four-cylinder piston engine (90 hp)
Wing span: 11 ft 6 in (3.50 m)
Length overall: 12 ft 7½ in (3.85 m)
Height overall: 2 ft 6 in (0.76 m)
Wing area: 18.7 sq ft (1.74 m^2)
Weight empty: 271 lb (123 kg)
Max launching weight: 360 lb (163 kg)
Max level speed: 194 knots (223 mph; 359 km/h) at S/L
Max rate of climb at S/L: 4,400 ft (1,341 m)/min
Service ceiling: 27,000 ft (8,230 m)
Range at S/L with max fuel: 183 nm (211 miles; 339 km)
Launch and recovery: Surface launch only, from land or ship either by rotary or zero launcher. Recovery by parachute released by radio command. Engine is stopped and parachute deploys automatically in the event of serious damage by gunfire or loss of radio control or electrical power.
Guidance and control: AN/ARW-79 radio command guidance system, with automatic altitude hold. Visual or radar tracking (radar or FM type tracking systems or equivalent).
Special equipment: 28V battery for all electrical power. L-band tracking system, smoke generating, tow banner and many other accessories available to meet requirements of individual users. For radar appearance augmentation, two wingtip reflector pods are optional.

Left: *MQM-36 in US Navy markings.* Right: *Shelduck built in Japan by Nippon Electric Company*

Remarks: Designed in 1946 and built originally for US Air Force and Army as MQM-33 (formerly OQ-19) and for US Navy as MQM-36 (formerly KD2R-5); current KD2R-5 is the export version of the MQM-36, used by the armed forces of 19 countries as a training target for ground-to-air gunnery and for surface-to-air and air-to-air missiles such as Seacat, Tigercat, Redeye, Blowpipe, Sparrow, Chaparral, Hawk, Sidewinder and Nike. More than 55,000 of all versions built since 1947, including 242 Shelducks by NEC in Japan for the JGSDF and JMSDF.

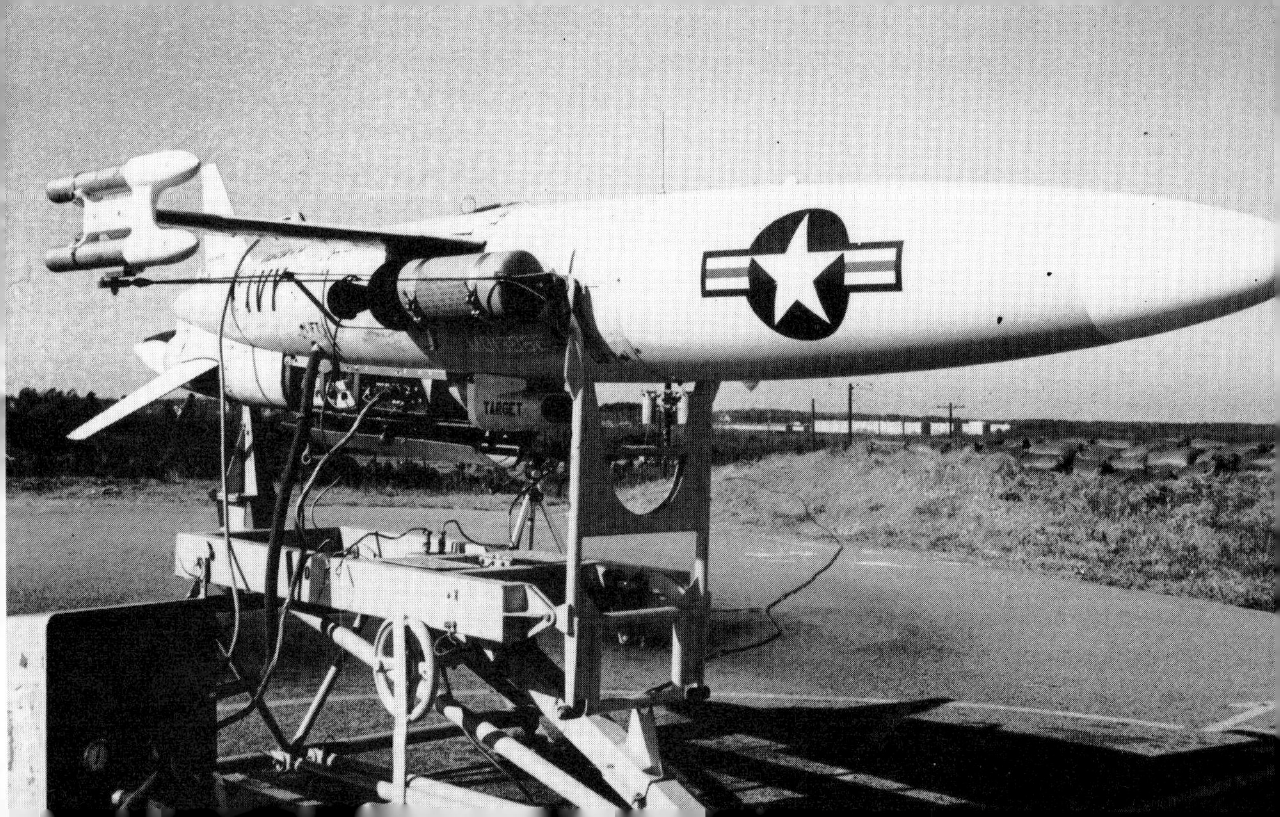
TARGET

Target drone; first flight 1967 (?)
Power plant: One Williams Research Corporation WR24-6 (YJ400-WR-400) turbojet engine (121 lb; 55 kg st)
Wing span: 5 ft 6 in (1.69 m)
Length overall: 11 ft 9 in (3.58 m)
Body diameter: 1 ft 2 in (0.36 m)
Height overall: 2 ft 3 in (0.70 m)
Weight empty: 233 lb (106 kg)
Max launching weight: 425 lb (192 kg)
Max level speed: 425 knots (489 mph; 787 km/h) at 20,000 ft (6,100 m)
Max rate of climb at S/L: 5,500 ft (1,676 m)/min
Service ceiling: 40,000 ft (12,200 m)
Range at max speed at 20,000 ft (6,100 m): 237 nm (273 miles; 439 km)
Launch and recovery: Zero-length launching by means of two Mk 34 Mod 1 JATO rockets and a ZL-5 launcher. Normal recovery by automatic drone pull-up followed by main parachute deployment, initiated automatically in emergencies such as interruption of continuous radio signal or loss of parachute command channel. Alternative method consists of direct main parachute deployment, and is initiated automatically on loss of electrical power. Main parachute, housed in fuselage immediately aft of wing, is a 30 ft (9.15 m) diameter extended-skirt nylon canopy, with automatic disconnect at impact.
Guidance and control: Radio command guidance system. Out-of-sight control by automatic stabilisation and command, with radar tracking; in-sight control with visual acquisition aids. Proportional feedback stabilisation and control system for pitch and bank. Engine throttle position, altitude hold initiation and recovery system initiation controlled by audio tone signals. Components include receiver-decoder, control unit assembly and vertical gyro in nose of fuselage, aileron and elevator servos, and altitude hold pressure transducer. Command control antenna in upper forward fuselage.
Special equipment: On-board acquisition and tracking aids include fore and aft Luneberg lenses for passive radar augmentation, four wingtip-mounted Mk 28 Mod 3 infra-red flares, active L-band augmentation, and a chaff or a pyrotechnic flash and smoke system designed to improve visual augmentation. Main payload compartment is in front fuselage between control equipment bay and fuel tank.
Remarks: Small, lightweight target for anti-aircraft gunnery, surface-to-air and air-to-air missile training, and weapon systems evaluation. Can provide training for the crews of 40 mm, 20 mm and 0.50 in guns, or Hawk, Redeye, Chaparral, Nike-Hercules, Sparrow and Sidewinder missiles. A modified version, with special augmentation, is used as a standard target at the NATO Missile Firing Installation (NAMFI) in the Mediterranean. The Italian Meteor Gufo tactical reconnaissance system (which see) is based on the Chukar I. A total of 2,300 MQM-74A/Chukar Is were built before production ended in 1973; these models have now been replaced in production by the MQM-74C/Chukar II (which see).

The **NV-128** was an experimental surveillance RPV, two of which were modified from MQM-74As and evaluated in 1974 (first flight in May) in the tri-service Solid Shield amphibious landing exercises. Overall length increased to 13 ft 4 in (4.08 m); max speed 450 knots (518 mph; 834 km/h); ceiling more than 35,000 ft (10,670 m). Higher-weight payload included CAI 70 mm camera as primary sensor, and forward oblique motion camera. Infra-red line-scanner

MQM-74A in US Navy service

or panoramic camera interchangeable with still camera. NV-128 programme is no longer active.

Testing began in latter part of 1975, at Pacific Missile Test Center, of XBQM-108A combining components of an MQM-74A and the engine and midcourse guidance system from a Harpoon missile. Tethered tests were to be followed by flight trials from a runway. (See entry under US Navy).

NORTHROP MQM-74C/CHUKAR II

Target drone; first flight 1973
Power plant: One Williams Research Corporation WR24-7 (J400-WR-401) turbojet engine (180 lb; 82 kg st)
Wing span: 5 ft 8 in (1.73 m)
Length overall: 12 ft 8½ in (3.87 m)
Body diameter: 1 ft 2 in (0.36 m)
Height overall: 2 ft 4 in (0.71 m)
Weight empty: 283 lb (128 kg)
Max launching weight: 492 lb (223 kg)
Max level speed: 515 knots (593 mph; 954 km/h) at 20,000 ft (6,100 m)
Max rate of climb at S/L: 5,840 ft (1,780 m)/min with full fuel
Service ceiling: 40,000 ft (12,200 m)
Range at max speed at 20,000 ft (6,100 m): 330 nm (380 miles; 611 km)
Launch and recovery: As for MQM-74A/Chukar I, except that two Mk 91 Mod 0 JATO rockets are used for launch.
Guidance and control: Generally as for MQM-74A/Chukar I, but with Motorola AN/DKW-1 integrated target control system. Chukar II has a developed control system using the MQM-74C autopilot instead of the MQM-74A decoder.
Special equipment: MQM-74C as for MQM-74A/Chukar I, plus pyrotechnic infra-red plume augmentors. Chukar II has Northrop Aerial Target Tracking System (NATTS), oil smoke system for visual augmentation, and wingtip-mounted radar augmented subtargets (RAST) system, but is otherwise similar to MQM-74C.
Remarks: Improved version of MQM-74A/Chukar I, with higher power and performance. In production for US Navy (718 MQM-74Cs) and for Imperial Iranian Navy (Chukar II); more than 700 delivered to US Navy by Spring 1977. Modified MQM-74Cs have been tested as RPVs, fitted with a Motorola radiation seeker and a warhead, under the US Navy's Persistent Arm programme. Three other MQM-74Cs adapted as NV-130 Nutcracker flight demonstration vehicles in US Air Force TEDS programme (which see).

MQM-74A (far left) *and MQM-74C* (left) *target drones*

RCS FALCON (UK)

Falcon mini-RPV, developed from the Merlin

Miniature target drone
Power plant: One piston engine (10 hp)
Wing span: 7 ft 0 in (2.13 m)
Length overall: 6 ft 6 in (1.98 m)
Payload: approx 10 lb (4.5 kg)
Max launching weight: 36 lb (16.3 kg)
Max speed: 110 knots (126.5 mph; 204 km/h)
Max controllable range (optical): 2.7 nm (3.1 miles; 5 km)
Endurance: 45 min
Launch and recovery: As for Merlin (which see).
Guidance and control: As for Merlin.
Special equipment: RCS4012 can carry X- and S-band radar augmentation devices, and can tow a 10 ft (3.05 m) drogue target on a 500 ft (150 m) cable. RCS4015 equipment same as for RCS4019 version of Merlin.
Variants:
RCS4012. Target system, for training air defence weapon crews to acquire and track low-level high-speed aircraft. Supplied to Libyan government in 1973-74.
RCS4015. Camera system, for short-range aerial reconnaissance. Target location by photo/map comparison. Supplied to Libyan government in 1973-74.
RCS4017. Composite target/camera system, combining facilities of RCS4012 and 4015.

RCS HERON and HERON HS

Miniature target drones
Power plant: One piston engine (10 hp)
Wing span:
Heron 9 ft 10 in (3.00 m)
Heron HS 8 ft 0 in (2.44 m)
Length overall: 7 ft 0 in (2.13 m)
Max payload:
Heron, catapult launch 15 lb (6.8 kg)
Heron, ground T-O 30 lb (13.6 kg)
Heron HS, catapult launch 10 lb (4.5 kg)
Max weight:
Heron, catapult launch 34 lb (15.4 kg)
Heron, ground T-O 53 lb (24 kg)
Max speed:
Heron 104 knots (120 mph; 193 km/h)
Heron HS 121.5 knots (140 mph; 225 km/h)
Max controllable range (optical): 2.7 nm (3.1 miles; 5 km)
Endurance: 30 min
Launch and recovery: As for Merlin (which see).
Guidance and control: As for Merlin.
Special equipment: X-band radar for tracking. Can carry up to 11 lb (5 kg) of underwing stores such as flares or small bombs.
Variants:
RCS4024 Heron. Designed for target work, but suitable for other applications. Ordered by British Army.
Heron HS. High-speed version, with reduced wing thickness and area, linear taper, and deletion of struts and pylons. Wheeled landing gear replaced by sprung metal skid and catapult launching points, although take-off dolly can be provided, permitting 25% increase in payload.

Photographs overleaf

Left: *RCS4024 Heron, in service with the British Army*

Above: *High-speed Heron HS in flight*

(UK) RCS MERLIN

Miniature target drone
Power plant: One 10 cc piston engine
Wing span: 4 ft 9 in (1.45 m)
Length overall: 4 ft 0 in (1.22 m)
Payload: 3 lb (1.4 kg)
Max launching weight: 10.5 lb (4.8 kg)
Max speed: 78 knots (90 mph; 145 km/h)
Max controllable range (optical): 1.6 nm (1.9 miles; 3 km)
Endurance: 30 min
Launch and recovery: Conventional take-off and landing, using non-retractable tricycle landing gear. Can also be used with system launcher.
Guidance and control: Six-function pulse width modulated VHF guidance system (RF output 5W).
Special equipment: RCS4018 can carry X- and S-band radar augmentation devices, for surveillance radar response, and can tow a 7 ft (2.13 m) drogue target on a 500 ft (150 m) cable for live firing target work. RCS4019 has as standard a radio-triggered Robot f/2.8 35 mm camera, giving 18 high-definition photographs per loading at altitudes up to 3,000 ft (915 m). Optional accessory kit for processing film and printing in approx 5 min after removal.
Variants:
RCS4018. Target system, for training air defence weapon crews to acquire and track target aircraft.
RCS4019. Camera system, for short-range aerial reconnaissance.
RCS4020. Composite target/camera system, combining facilities of RCS4018 and 4019.

RCS4020 Merlin composite target/camera system

RCS MOSSETTE

(UK)

Miniature target drone
Power plant: One 6 cc piston engine
Wing span: 4 ft 0 in (1.22 m)
Length overall: 3 ft 6 in (1.07 m)
Max launching weight: 4 lb (1.8 kg)
Max speed: 69 knots (80 mph; 128 km/h)
Max controllable range (optical): 1.6 nm (1.9 miles; 3 km)
Endurance: 20 min
Launch and recovery: Launch from system catapult; net recovery system.
Guidance and control: As for Merlin (which see).
Special equipment: Smokeless flare beneath underfin to aid target alignment.
Variant:
RCS4041. Re-usable target, with in-flight profile similar to current wire-guided anti-tank missiles; intended for training operators of helicopter- and vehicle-launched missiles of that type. Expendable nose-cone.

Mossette target on its launcher

Training drone
Power plant: One 10 cc piston engine
Wing span: 7 ft 0 in (2.13 m)
Length overall: 6 ft 0 in (1.83 m)
Payload: 4 lb (1.8 kg)
Weight dry: 11 lb (5 kg)
Max speed: 52 knots (60 mph; 96.5 km/h)
Endurance: 35 min
Launch and recovery: As for Merlin (which see).
Guidance and control: Six-channel proportional radio control, designed for use with standard tripod-mounted Merlin system.
Variant:
RCS4030. For training operators of operational drone/RPV systems.

RCS4030 training drone

RCS SATS

(UK)

Miniature target drone
Wing span: 7 ft 3 in (2.21 m)
Length overall: 6 ft 6 in (1.98 m)
T-O weight: 9 lb (4.1 kg)
Max speed: 74 knots (85 mph; 137 km/h)
Endurance: 40 min
Guidance and control: Radio control from hand-held transmitter.
Remarks: SATS (Small Arms Target System) is used as a live target for infantry anti-aircraft firing practice with rifles and sub-machine-guns.

RCS Small Arms Target System (SATS)

RCS SWIFT

Reconnaissance or target drone
Wing span: 6 ft 0 in (1.83 m)
Payload: 2 lb (0.9 kg)
Max T-O weight: 10 lb (4.5 kg)
Max speed: 100-120 knots (115-138 mph; 185-222 km/h) according to engine fitted
Endurance: 30 min
Range: 4.3 nm (5 miles; 8 km)
Launch and recovery: Launched from specially designed toggle action catapult, and flown along fixed heading at fixed angle of climb, using optical tracking system.
Guidance and control: Basic system includes tripod-mounted high-power radio transmitter (range 5.4 nm; 6.2 miles; 10 km) and optical sight. Drone can also be operated with six-channel hand-held transmitter.
Special equipment: 35 mm camera, barometric trip switch and marker flash.
Remarks: Glassfibre fuselage; swept wings of plywood-covered styrofoam core.

Swift sweptwing photo-reconnaissance or target drone on its catapult launcher

RCS SMALL ROTARY-WING DRONE (UK)

Miniature target drone
Power plant: One 10 cc piston engine (1.2 hp)
Main rotor diameter: 5 ft 4 in (1.63 m)
Tail rotor diameter: 1 ft 2 in (0.36 m)
Length overall: 6 ft 0 in (1.83 m)
Payload: 2 lb (0.9 kg)
Weight dry: 11 lb (5 kg)
Max speed: 61 knots (70 mph; 113 km/h)
Endurance: 20 min
Launch and recovery: Conventional helicopter take-off and landing, using non-retractable tricycle landing gear.
Guidance and control: Six-channel proportional radio control system (four for pitch, roll, yaw and throttle control, and two for camera controls).
Special equipment: Choice of half-frame (f/2.8) or full-frame (f/2.5) 35 mm camera, each with 18 exposures per wind and automatic film advance. Heavy-duty 12V electric starter optional.
Remarks: Intended primarily as target for training operators of radar-controlled missiles.

RCS mini-helicopter drone

RCS LARGE ROTARY-WING RPV

Surveillance mini-RPV
Power plant: One two-stroke piston engine (10 hp)
Main rotor diameter: 7 ft 6 in (2.29 m)
Payload: 18.7 lb (8.5 kg)
Max T-O weight: 44 lb (20 kg)
Max speed: 32 knots (37 mph; 60 km/h)
Guidance radius: 2.7 nm (3.1 miles; 5 km)
Launch and recovery: Conventional helicopter take-off and landing, using wide-track tubular skid landing gear.
Special equipment: Standard payload comprises ground surveillance TV system, cameras, sensors and other equipment. Gyroscopic stabilisation optional.
Remarks: Twin-engined version, with 40 lb (18 kg) payload capacity, under development.

Single-engined version of RCS's large rotary-wing RPV, intended for surveillance duties

HiMAT RPRV
NASA

ROCKWELL INTERNATIONAL/NASA HiMAT

Remotely piloted research vehicle; first flight due early 1978
Power plant: One General Electric J85-GE-21 turbojet engine (3,500 lb; 1,588 kg st dry, 5,000 lb; 2,268 kg st with afterburning)
Wing span: 15 ft 7.2 in (4.75 m)
Length overall, incl probe: 22 ft 6 in (6.86 m)
Height overall: 4 ft 3.6 in (1.31 m)
Weight empty: 2,800 lb (1,270 kg)
Max air-launching weight: 3,400 lb (1,542 kg)
Max level speed (estimated): Mach 1.4 for 3 min at 40,000 ft (12,200 m)
Launch and recovery: Air-launched from B-52 carrier aircraft at approx 45,000 ft (13,720 m). Recovery by conventional landing on retractable tricycle-type skid landing gear on dry lake bed at Edwards AFB.
Guidance and control: Primary control from ground console, by TV, telemetry and radar link with onboard systems. If ground control is lost, backup inputs from the RPRV will be relayed to a TF-104G chase-plane. On occasions when the TF-104G is outside of control range, the HiMAT has an onboard self-righting system that will bring it into straight and level subsonic flight until the TF-104G can resume control.
Special equipment: Onboard equipment includes TV camera in cockpit, radar altimeter under nose and angle of attack sensor in nose probe.
Remarks: NASA programme to speed up the progress of advanced design technology into the flight test phase; to assist designers in taking larger technological steps forward between generations of aircraft; and to provide a low-cost, low-risk means of testing the advanced manoeuvring capability of future aircraft. (HiMAT is an abbreviation for Highly Manoeuvrable Aircraft Technology.) Blended wing/body design will permit exchange of modular components to test such different features as supercritical or variable-camber wings, digital fly-by-wire systems, and vectored-thrust or variable-thrust engine systems. Flight evaluation to begin in early 1978, using two Rockwell-built prototypes, and to last for approx 1½ years.

Left: *Artist's impression of HiMAT, in its original configuration, being air-launched from a B-52 Stratofortress.* Right: *Model showing current HiMAT configuration*

LONG
VEHICLE
73BM06

Small target drone; first flight 1975 (?)
Power plant: One 123 cc single-cylinder piston engine (12.5 hp)
Wing span: 11 ft 0 in (3.35 m)
Length overall: 8 ft 0 in (2.44 m)
Payload: 15 lb (6.8 kg)
Max launching weight: 100 lb (45.4 kg)
Max level speed: 125 knots (144 mph; 231 km/h)
Min endurance: 50 min (at max speed)
Launch and recovery: Launched from pneumatically-operated vehicle-mounted rail launcher. Can be launched into headwinds of up to 25 knots (29 mph; 46 km/h). Recovery by triple parachute, or by conventional landing using the flares housing as a skid.
Guidance and control: Seven-channel radio command guidance system (hand-held transmitter), in 68MHz band. Aerodynamic control by ailerons and all-moving tailplane, all servo-assisted. Roll stabilisation (from sensor in port wing near tip) and automatic (barometric) height control standard.
Special equipment: Tracking flares or smoke generators in underfuselage payload compartment. Radar reflector in centre of fuselage. Recovery parachute in compartment in nose.
Remarks: MATS (Model Aircraft Target System) was developed under MoD contract for use in practice firings of Shorts Blowpipe, other small anti-aircraft weapons, and AA artillery. Construction is of metal and glassfibre-reinforced plastics, with foam-filled wings, fin and tailplane.

Left: *MATS-B leaving its launcher.* Right: *Basic MATS-B vehicle without underbelly payload pack*

Skysp
lowpipe

VTOL tactical RPV; first (tethered) flight 1975
Data apply to prototype
Power plant: One two-stroke piston engine (65 hp), driving a multi-blade ducted fan
Overall diameter: 3 ft 6½ in (1.08 m)
Fan diameter: 2 ft 9½ in (0.85 m)
Height overall: 4 ft 6 in (1.37 m)
Weight empty: 187 lb (85 kg)
Payload: 44 lb (20 kg)
Max T-O weight: 286 lb (130 kg)
Max speed: 102 knots (118 mph; 190 km/h)
Max operating altitude: 6,000 ft (1,825 m)
Endurance with 44 lb (20 kg) payload: 1 hr 30 min
Launch and recovery: Vertical take-off and landing. Launching arm supports vehicle in T-O attitude until required thrust for VTO is achieved. Free vertical landings can be made on level ground; landing net available for recovery in rough terrain.
Guidance and control: Radio command guidance system. Aerodynamic control by aerofoil surfaces set across the duct exit and integrated in part with low aspect ratio wings on the exterior of the duct. Lift and propulsive thrust provided by vectoring the gross thrust of the ducted fan power plant.
Special equipment: Autopilot and power supply equipment at rear of vehicle, in wall of duct. Operational equipment, variable according to mission, can include TV camera, sensors, automatic data-gathering, laser and other equipment; this is installed in pod fairing on outer forward surface of the duct. Skyspy is intended to be flown under remote control to the chosen surveillance area, where it can hover over a stationary target (or track a moving one) and relay positional details of the target in real time, using secure data links, to a ground- or ship-based controller.
Remarks: Design began in 1972. Currently under development for a variety of potential missions, including army reconnaissance; naval over-the-horizon viewing; weapon control and delivery, including ability to provide command link for over-the-horizon weapon control systems; electronic countermeasures; target spotting; coastguard surveillance; border patrol and police duties; fishery protection; search and rescue; forest fire spotting; and emergency relief and medical support service. Surveillance can be carried out at all angles of attack between conventional forward flight and the hover mode.

Skyspy surveillance and general purpose RPV

U.S. AIR FORCE
0-61419

Target drone; first flight 13 August 1974
Power plant: One Pratt & Whitney J57-P-23A turbojet engine (11,700 lb; 5,307 kg st dry, 17,200 lb; 7,802 kg st with afterburning)
Wing span: 38 ft 1½ in (11.62 m)
Length overall: 68 ft 4⅔ in (20.84 m)
Height overall: 21 ft 2½ in (6.46 m)
Wing area: 695.0 sq ft (64.57 m²)
Max level speed: 688 knots (792 mph; 1,274 km/h) (Mach 1.2)
Operating height range: 200 ft (61 m) to 55,000 ft (16,760 m)
Launch and recovery: Conventional take-off and landing, using retractable tricycle landing gear.
Guidance and control: Ground controlled through a remote data link. Pre-programmed manoeuvring capability, which can be initiated or terminated by ground controller. Vega Precision Labs makes certain control electronics under subcontract.
Special equipment: No details known.
Variants:
QF-102A. Version retaining normal cockpit controls and provision for onboard check pilot. Initial order for two (later increased to five).
PQM-102. Major production version, designed and equipped entirely for unmanned operation. First-ever fighter aircraft converted for drone duties with no provision whatever for manned flight, and cannot be flown except under remote control. Initial order for six (since increased to 76), of a planned total USAF procurement of 128 (including QF-102As) over a six-year period. First operational flight 25 June 1975. The US Army will receive 24 of the first 68 for ground-to-air missile evaluation.
Remarks: Converted by Sperry Flight Systems Division from former Convair F-102A Delta Dagger interceptors to provide up-to-date "threat simulation" targets for US Air Force air-to-air weapon tests. Has greater speed and manoeuvrability than the manned F-102A resulting from lower operating weight. Airframe modification carried out by Aircraft Service Division of Fairchild Industries, which has also developed a prototype of a less expensive PQM-102 for future conversions.

Left: *PQM-102, the first converted-fighter target drone to have no provision for manned operation.* Overleaf: *PQM-102 landing under remote control from the van in the foreground (left) and control desks for the PQM-102 inside the control van (right)*

U.S. AIR FORCE

TELEDYNE RYAN MODEL 124 (BQM-34/MQM-34) FIREBEE I

Subsonic target drone; first flight Summer 1951
Data apply to BQM-34A
Power plant: One Teledyne CAE J69-T-29 turbojet engine (1,700 lb; 771 kg st)
Wing span: 12 ft 10.8 in (3.93 m)
Length overall: 22 ft 10.8 in (6.98 m)
Body diameter: 3 ft 1.2 in (0.94 m)
Height overall: 6 ft 8.4 in (2.04 m)
Wing area: 36.0 sq ft (3.34 m^2)
Weight empty: 1,500 lb (680 kg)
Max launching weight: 2,500 lb (1,134 kg)
Max level speed: 600 knots (690 mph; 1,112 km/h) at 6,500 ft (1,980 m)
Max rate of climb at S/L: 16,000 ft (4,875 m)/min at 2,200 lb (1,000 kg) AUW
Operating height range: 50 ft (15 m) to more than 60,000 ft (18,300 m)
Max range: 692 nm (796 miles; 1,282 km)
Launch and recovery: Either air-launching, from a suitably-modified aircraft, or ground-launching, using an 11,300 lb (5,125 kg) st (nominal) solid-propellant JATO bottle. The US Navy has also launched BQM-34As from ships under way at up to 15 knots (17 mph; 27.5 km/h). Two-stage parachute recovery system, operating automatically in the event of a target hit, loss of radio wave carrier from the remote control station, engine failure, or upon command by the remote control operator. To prevent damage by dragging, recovery system incorporates a disconnect which releases the parachute from the drone on contact with surface.

Firebee I launch at the US Navy's South Atlantic/Caribbean Sea Weapons Range, Cabras Island, Puerto Rico

Guidance and control: Remote control methods include choice of radar, radio, active seeker and automatic navigator. Normal method is through a UHF radio link using an AN/FRW-2 or SRW-4 ground transmitter and an AN/DRW-29 airborne receiver. Target can be controlled either from a manned aircraft or from a surface station. Remote command includes activation of special scoring and augmentation equipment in the target. An onboard beacon facilitates radar tracking, and provision exists for telemetry system if required. Basic commands consist primarily of on/off functions, received by the onboard radio receiver and relayed to the appropriate subsystem. A Motorola ITCS (integrated tracking and control system) is fitted in the BQM-34S version. Other types of remote command and tracking systems can include a microwave command and guidance system to control the Firebee beyond line-of-sight from a ground station through an airborne relay station. This operates with coded impulses which reduce possible interference from other electronic signals. Operational Firebees can be equipped with an increased manoeuvrability three-axis flight control system for tactical air combat simulation which gives the target the capability to perform 4, 5 or 6*g* manoeuvres. A Radar Altimeter Low Altitude Control System (RALACS), when added to the Firebee I control system, permits precision low-altitude flights at 50 ft (15 m) over water and 100 ft (30 m) over land.
Special equipment: AN/DRW-29 radio receiver with Dorsett TM-4-31A telemetry system, or DKW-2 guidance transponder. A/A37G-3 or A/A37G-8 flight control system. Wide range of "building block" operational equipment includes visual or radar-reflecting banner targets; radar or infra-red Towbee towed targets or tow target Doppler "bird"; two underwing drop-tanks, 500 lb bombs or bomblet dispensers; AN/ALE-33 or other ECM containers;

Above: *MQM-34D Firebee I with wingtip-mounted Towbee infra-red towed target.* Right: *Launch of Towbee-equipped MQM-34D.* Opposite page: *Four Firebee I targets under the wings of a Lockheed DC-130 Hercules launch/director aircraft*

wingtip tow launchers, camera pods, scoring equipment, flares or other forms of infra-red augmentation; or reflector pods for radar augmentation. The BQM-34A can be equipped with adjustable travelling wave tube amplifiers for use as radar echo enhancers in the L, S, X and C frequency bands.

Variants:

BQM-34A. Standard USAF/US Navy version, as described. Developed from earlier Q-2A/KDA series and formerly known as Q-2C. Total of more than 4,480 BQM-34A/Ss and MQM-34Ds built by February 1977, since production began in 1959.

BQM-34S. Designation of US Navy Firebee Is fitted with Motorola ITCS or Vega track and control system (VTCS). In production.

MQM-34D. Designation of US Army version, which has extended-span wings and a longer-burning rocket booster, enabling it to take off at a loaded weight some 1,000 lb (455 kg) greater than BQM-34A. Prototypes also built of an MQM-34D/MOD II (Teledyne Ryan Model 251), but this version not proceeded with.

Remarks: Total production, to February 1977, includes 1,280 of the earlier Q-2A/KDA versions for the three US services and the RCAF; and 5,815 of current-model Firebee Is, including 21 being built under licence by Fuji for the Japan Defence Agency. The Model 124/Firebee I airframe is also the basis for later Ryan-developed Models 147, 234, 255 and 259 RPVs (which see).

US Navy Ship Research and Development Center began in 1975 to modify a Firebee into a VTOL RPV by fitting delta-planform canard surfaces and thrust deflection vanes to the tailplane.

Right: *BQM-34A dropping away from the launch pylon of its director aircraft.* Opposite page: *Launching (left) and controlling (right) a BQM-34A Firebee I target of the USAF, from a DC-130 Hercules*

U.S. AIR FORCE

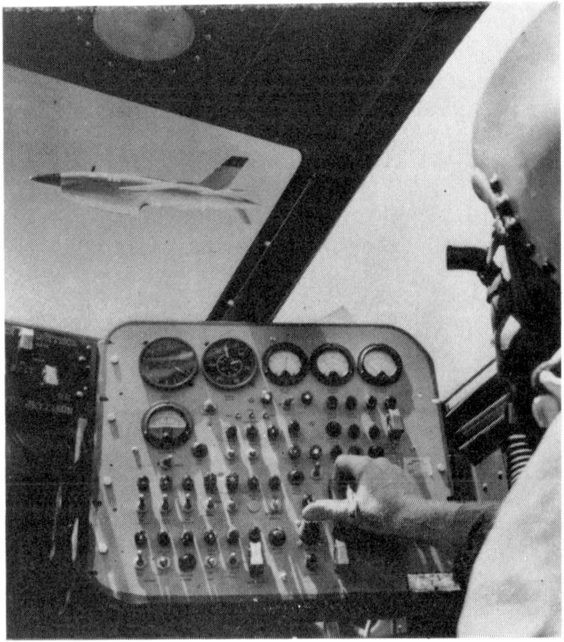

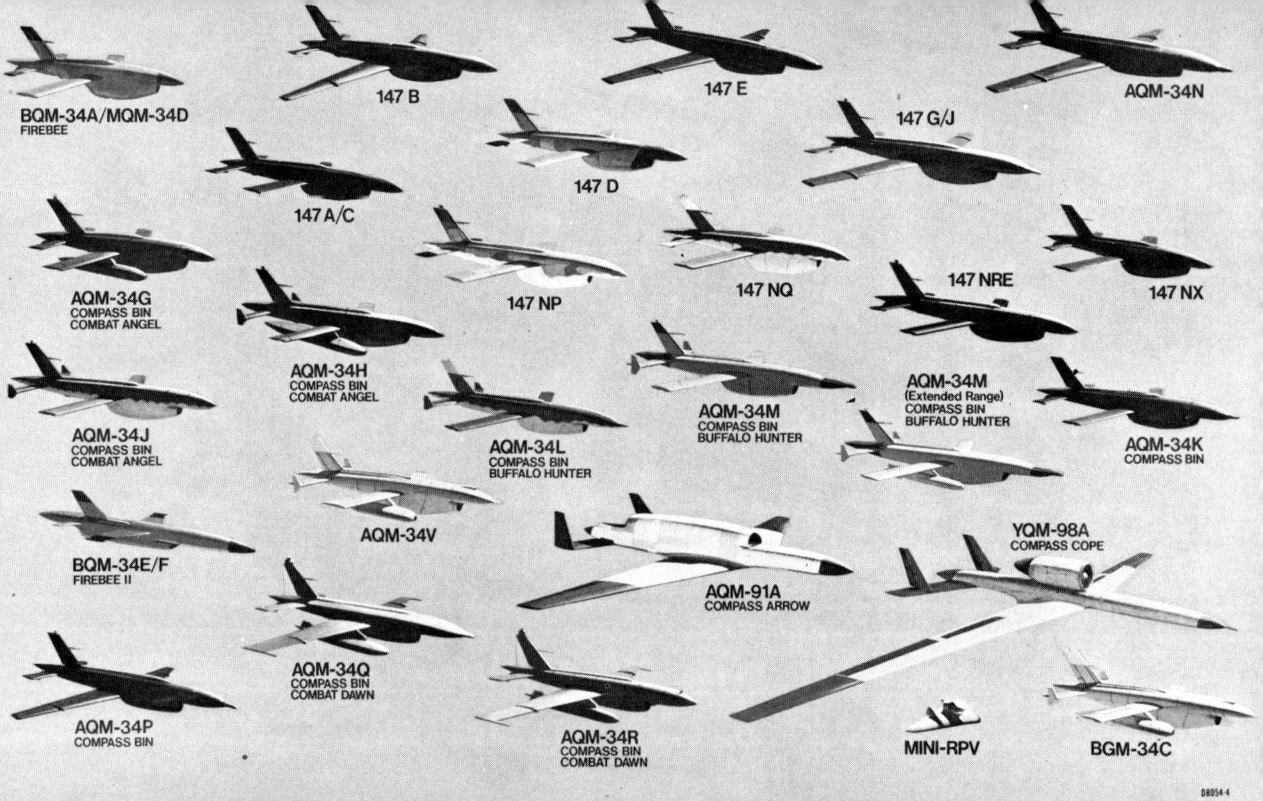
BQM-34A/MQM-34D
FIREBEE
147 B
147 E
AQM-34N
147 G/J
147 A/C
147 D
AQM-34G
COMPASS BIN
COMBAT ANGEL
147 NP
147 NQ
147 NRE
147 NX
AQM-34H
COMPASS BIN
COMBAT ANGEL
AQM-34M
(Extended Range)
COMPASS BIN
BUFFALO HUNTER
AQM-34J
COMPASS BIN
COMBAT ANGEL
AQM-34L
COMPASS BIN
BUFFALO HUNTER
AQM-34M
COMPASS BIN
BUFFALO HUNTER
AQM-34K
COMPASS BIN
AQM-34V
BQM-34E/F
FIREBEE II
YQM-98A
COMPASS COPE
AQM-91A
COMPASS ARROW
AQM-34Q
COMPASS BIN
COMBAT DAWN
AQM-34P
COMPASS BIN
AQM-34R
COMPASS BIN
COMBAT DAWN
MINI-RPV
BGM-34C
08054-4

TELEDYNE RYAN MODEL 147 and 255 series (AQM-34/XQM-103)

Tactical RPVs for reconnaissance, ECM, surveillance and training; first flight 1962
Power plant: One Teledyne CAE J69-T-29 turbojet engine (1,700 lb; 771 kg st) in Models 147NA, NC, NC(M-1) and 255;
or: One Teledyne CAE J69-T-41A turbojet (1,920 lb; 871 kg st) in Models 124I, 147H, SC, SD and SRE;
or: One Teledyne CAE J100-CA-100 turbojet (2,700 lb; 1,225 kg st) in Models 147T, TE and TF
Wing span:
147SC 13 ft 0 in (3.96 m)
124I, 147NA, NC, NC(M-1), SD, SRE, 255 14 ft 6 in (4.42 m)
147H, T, TE, TF 32 ft 0 in (9.75 m)
Length overall:
147NA, NC, NC(M-1), 255 26 ft 0 in (7.92 m)
147SRE 29 ft 0 in (8.84 m)
147H, SC, SD, T, TE, TF 30 ft 0 in (9.14 m)
124I 31 ft 0 in (9.45 m)
Body diameter:
124I, 147H, NA, NC, NC(M-1), SC, SD, SRE, 255 3 ft 1.2 in (0.94 m)
147T, TE, TF 3 ft 3.6 in (1.01 m)
Max launching weight:
124I 3,250 lb (1,474 kg)
147H 3,820 lb (1,732 kg)
147NA 3,684 lb (1,671 kg)
147NC (AQM-34H) 3,749 lb (1,700 kg)
147NC(M-1) 2,865 lb (1,299 kg)
147SC 3,065 lb (1,390 kg)
147SC(TV) 3,150 lb (1,429 kg)

Currently-identifiable drones and RPVs designed by Teledyne Ryan, showing USAF designations and programme associations

147SD, SD Loran 3,113 lb (1,412 kg)
147SRE 3,367 lb (1,527 kg)
147T 3,792 lb (1,720 kg)
147TE 3,870 lb (1,755 kg)
147TF:
without underwing stores 4,100 lb (1,859 kg)
with underwing stores 6,200 lb (2,812 kg)
225 (AQM-34V) 4,500 lb (2,041 kg)
Cruising speed (147TF): 420 knots (485 mph; 780 km/h) above 60,000 ft (18,300 m)
Launch and recovery: All models normally air-launched from DC-130A, E or H Hercules carrier/director aircraft and recovered by CH-53 helicopter mid-air retrieval system (MARS). AQM-34V tested in ground launch in 1975 using Thiokol booster rocket and TRW thrust vector control.
Guidance and control: Teledyne Ryan radio command; or programme and command; or radio/microwave command guidance system. YAQM-34U (Update) vehicles fitted with new modular automatic flight control system; microwave command guidance and attitude heading reference subsystems; Doppler radar; Loran receiver; radar altimeter (also in AQM-34M); digital computer; recovery control subsystem; and aileron/elevator/rudder servos. Loran receiver also fitted to AQM-34M(L). 147G (piloted from FDL-23 ground remote control console) fitted with Cohu TV camera, Aacom 20W L-band TV transmitter, and modified three-axis proportional control system. AQM-34H fitted with Sperry Univac APW-25 or -26 transponder.
Special equipment: AQM-34H can carry two underwing ECM (ALE-2 chaff dispenser) pods. Similar pods carried by AQM-34G. The AQM-34L carries Fairchild F415Y camera (2,000 exposures on 75 mm film) and TV relay; some also carry active/passive ECM pods

H I
423

Left: *Teledyne Ryan Model 147NC (AQM-34V) electronic warfare RPV.* Above: *Launching an AQM-34R (Model 147TF) from a DC-130E Hercules*

underwing. YAQM-34U (Update) vehicles had new modular avionics installation in rear fuselage; Fairchild camera and annotation system; and a pylon-mounted acoustic or other sensor pod under each wing. Compass Robin AQM-34M fitted with micro-electronic radio receivers, ejected near enemy radar to collect and relay elint data. AQM-34V has new E-systems modular noise jammers in nose and MB Associates ALE-38 underwing chaff dispenser pods.

Variants:

Model 147A. Original version, basically a Firebee I with new guidance system and more fuel. Developed 1962 as part of USAF Big Safari intelligence-gathering programme.

Model 147B. Lengthened fuselage (approx 29 ft; 8.8 m) and extended-span wings (approx 27 ft; 8.23 m).

Model 147C. Externally similar to 147A.

Model 147D. Externally similar to 147A.

Model 147E. Externally similar to 147B but with blunter, more bulbous nose and longer dorsal spine.

Model 147G. Externally similar to 147B but with narrow probe-like fairing on nose and small fin-tip antenna fairing. Two modified extensively under USAF Red Baron programme to obtain in-flight control and guidance data. Reportedly used also in Bell study for an air cushion landing system for RPV applications.

Modified Model 147G (XQM-103). Much modified 147G with structure strengthened to withstand 10*g*. For research into high-*g* manoeuvring; target acquisition; remote piloting techniques, including relationship between RPV and its remote pilot; and various subsystem evaluations. Programme dependent on future funding.

Model 147H (AQM-34N). Medium altitude reconnaissance version. Externally similar to 147G except for flatter ventral bulge

Left: *AQM-34V with underwing chaff dispensing pods.* Above: *Model 147SD (AQM-34M) low-altitude photo-reconnaissance RPV*

similar to that of 147T series. Became operational with Strategic Air Command in about 1968. No longer operational.

124I. Hybrid Model 124/147: about a dozen said to have been supplied to Israel in 1971 and later used for high altitude photo-reconnaissance overflights of Arab territory. Other unconfirmed reports suggested that Israel has used Firebee I-type drones (not necessarily of the same type) to delivery Israeli-built air-to-ground missiles against Egyptian missile sites. Model 124I, used in October 1973 war for day and night reconnaissance, has endplate auxiliary fins and elongated nose similar to USAF's AQM-34L and M. Israeli RPVs also used in 1973 war to decoy Arab SAM missiles, one successfully drawing the fire of 32 SAMs and still returning to its base.

Model 147J. Externally similar to 147G.

Model 147NA (AQM-34G). Medium altitude ECM version. Extended-span wings, strengthened to carry active (jamming) or passive (chaff dispensing) ECM pod under each wing. Small antenna fairing on top of fin. Built under USAF Compass Bin and Combat Angel programmes. No longer in service.

Model 147NC (AQM-34H). Medium altitude ECM version, similar to AQM-34G except for endplate auxiliary fins. Built under USAF Compass Bin and Combat Angel programmes. Developed for use in Vietnam, but not used operationally. Entered service from 1969 with 11th Tactical Drone Squadron. Some updated to AQM-34V. No longer in service.

Model 147NC (M-1) (AQM-34J). Medium altitude training version of AQM-34H, without underwing pylons. Operated for reconnaissance RPV training with 11th Tactical Drone Squadron. Some updated to AQM-34V. No longer in service.

Model 147NP. Externally similar to AQM-34J, but without auxiliary fins.

Tom Cat

Left: *Each symbol painted on the side of this AQM-34L records a completed operational mission.* Above: *AQM-34H with ALE-2 underwing chaff-dispensing pods*

Model 147NQ. Externally similar to 147NP.
Model 147NRE. Externally similar to 147NP.
Model 147NX. Externally similar to 147NQ but with shorter-span wings, no underwing pylons, no fin-tip antenna fairing.
Model 147SA. Externally similar to 147NP, but with probe-like nose fairing.
Model 147SB. Externally similar to 147SA.
Model 147SC (AQM-34L and YAQM-34U). Low altitude photo-reconnaissance versions. Externally similar to 147SA/SB. Built under USAF Compass Bin and Buffalo Hunter programmes. AQM-34L was only reconnaissance RPV permitted to overfly N Vietnam after cessation of bombing on 15 January 1973. Six AQM-34L modified to YAQM-34U by Lear Siegler in 1972 Update programme to improve low level navigational accuracy and reliability of RPV avionics system. Five YAQM-34U later converted as prototypes of multi-mission Model 259/BGM-34C (which see).
Model 147SD (AQM-34M). Low altitude photo-reconnaissance version, generally similar to AQM-34L; 87 built under USAF Compass Bin and Buffalo Hunter programmes. Some converted to **AQM-34M(L) (147SD Loran)**, and one as Compass Robin elint testbed. Extended-range version also built, with two underwing drop-tanks.
Model 147SK. Version evaluated by USN for "special purpose reconnaissance"; no details known. Damaged airframes displayed by China and N Vietnam reported to include some SKs.
Model 147SRE (AQM-34K). Low altitude night reconnaissance version. Similar to others in 147S series but with shorter fuselage and no fin-tip fairing. Built under USAF Compass Bin programme.
Model 147T (AQM-34P). High altitude surveillance version, built under USAF Compass Bin programme. Damaged airframes displayed by China and N Vietnam included a 147T.

Model 147TE (AQM-34Q). Medium altitude surveillance version, built under USAF Compass Bin and Combat Dawn programmes. Extended-span, broader-chord wings, with tubular weight fairings at tips, and two underwing drop-tanks. Bulbous data link fairing on top of fin.

Model 147TF (AQM-34R). High altitude surveillance version. Externally similar to 147TE, but with blade antenna above nose and no nose probe. Built under USAF Compass Bin and Combat Dawn programmes; 20 ordered in 1971. Provision for underwing stores.

Model 255 (AQM-34V). Updated version of AQM-34H and J, with increased active/passive ECM capability. Equipped with E-Systems (Melpar Division) active jamming equipment and carries either Lundy ALE-2 or MB Associates ALE-38 chaff dispenser pods on underwing hardpoint pylons. Air or ground launch capability; primary retrieval by MARS. Forty-seven converted from existing AQM-34H/J, and 16 ordered as new. Deliveries took place in 1976. In service with 432nd Tactical Drone Group.

Remarks: From late 1964 until the cease-fire in 1973, various types of Model 147 made flights over China, N Vietnam and other areas. Early production versions deployed initially from Okinawa as part of USAF Blue Springs programme for overflying Chinese mainland. In late 1960s, detachments of 100th Strategic Reconnaissance Wing operated 147-type RPVs from Bien Hoa in S Vietnam, recovering them by MARS from Da Nang air base. In all, USAF flew more than 2,500 RPV sorties in Southeast Asia, with outstanding results. Total production of Model 147 series runs to many hundreds. Production of 147A/B/C/D/E/G/J models has ended.

Far left: *AQM-34Q medium-altitude surveillance drone at an operational air base.* Left: *Thirteen 'hashmarks' painted on the fuselage of this AQM-34Q denote missions already completed*

TELEDYNE RYAN MODEL 154 (AQM-91A)

Left: *AQM-91A under the wing of its DC-130E launch/director aircraft.* Above: *After mid-air retrieval, an AQM-91A is carried home by a Sikorsky HH-3E recovery helicopter. The small drogue chute stabilises the drone in flight under the HH-3E*

High-altitude surveillance RPV; first flight 1968 (?)

Power plant: One General Electric J97-GE-3 turbojet engine (5,270 lb; 2,390 kg st)

Wing span: 48 ft 0 in (14.63 m)

Length overall: 34 ft 0 in (10.36 m)

Body diameter: 3 ft 2.4 in (0.975 m)

Max launching weight: 5,245 lb (2,379 kg)

Control range: approx 175 nm (200 miles; 320 km)

Launch and recovery: Air launch from beneath wings of DC-130E Hercules director aircraft. Parachute recovery, by helicopter mid-air retrieval system (MARS).

Guidance and control: Guidance to and from target area was by a programmed navigation system utilising a Teledyne Ryan Doppler sensor, inertial stabilised platform and digital computer. Final recovery by Sperry Univac UPQ-3 microwave command guidance system, operated either from recovery aircraft or from ground station via onboard transponders.

Special equipment: Basic reconnaissance sensor is an Itek KA-80A optical bar panoramic camera with a focal length of 24 in (610 mm). Weighing about 325 lb (147 kg) with a full magazine of film, this camera can take some 1,500 exposures, each 45 in (114 cm) long by 4.5 in (11.4 cm) wide, on each mission. Raytheon elint (electronic intelligence) and HRB Singer infra-red sensors also carried.

Remarks: Development began in 1966 as a high-altitude reconnaissance and electronic surveillance RPV for operation over China, under the USAF's now-inactive Compass Arrow programme. Overflights of mainland China were halted in mid-1971, and in July 1973 all AQM-91As were put into store. They remain available for reactivation if a requirement should arise.

NAVY
BEWAR
AND JET BLASTS

TELEDYNE RYAN MODEL 166 (BQM-34E/F/T) FIREBEE II

Supersonic target drone; first flight 1968
Data apply to BQM-34E
Power plant: One Teledyne CAE YJ69-T-406 turbojet engine (1,920 lb; 871 kg st); provision for solid-propellant JATO bottle (11,300 lb; 5,125 kg st nominal)
Wing span: 8 ft 10.8 in (2.71 m)
Length overall: 28 ft 8 in (8.73 m)
Body diameter: 2 ft 0 in (0.61 m)
Height overall: 5 ft 7.2 in (1.71 m)
Weight empty: 1,452 lb (658 kg)
Max launching weight:
air launch 1,886.5 lb (855.7 kg)
ground launch 2,264.5 lb (1,027.2 kg)
Max level speed: Mach 1.8 at 45,000 ft (13,720 m)
Service ceiling: 60,000 ft (18,300 m)
Max range: 774 nm (891 miles; 1,434 km)
Launch and recovery: Launched from ground or shipborne launcher, or from a modified DP-2E Neptune, DC-130A or DC-130E Hercules director aircraft (air launch at altitudes up to 18,000 ft; 5,485 m and at speeds of approx 200 knots; 230 mph; 370 km/h). Recovery by two-stage parachute system similar to that of Firebee I. Can also be recovered from water or (BQM-34F only) by helicopter mid-air retrieval system. In a MARS recovery, the helicopter snares an 18 ft 9 in (5.72 m) diameter engagement parachute which extends above the 79 ft 0 in (24.08 m) main chute, the main chute then being released automatically.
Guidance and control: Frequency-modulated UHF radio guidance system, with 20 separate command channels, utilising an AN/DRW-29 onboard radio receiver with a compatible transmitter at the remote control station. Six-element automatic flight control system comprises three-axis rate gyro, vertical gyro, air data computer, flight control box, low-altitude control box, and three-axis electro-hydraulic actuator assembly. Positioning data provided by onboard radar tracking beacon. Ten-channel telemetry system comprises data collection, conversion and FM/FM transmitting equipment in the drone and receiving and data display units in the remote control station. Flight control system continues to operate on battery power when engine or generator power is shut down at high altitude.
Special equipment: AN/DLQ-3 ECM equipment, AN/DRQ-4 missile scoring system, AN/APX-71 L-band beacon, special low-altitude radar altimeter kit for 50 ft (15 m) altitude, X- or C-band tracking beacons, AN/DRW-29 radio control receiver, Dorsett AN/AKT-21 telemetry system. (In BQM-34T, Motorola ITCS performs functions of tracking beacon, radio control receiver and telemetry system.) Radar augmentation includes travelling wave tube (TWT) in S-, C- and X-band and a nose-mounted Luneberg lens passive radar reflector. Infra-red augmentation by wingtip-mounted Mk 37 Mod 0 flares. In general, Firebee II can be equipped with active and passive radar augmentors, electronic and photographic scoring systems, ECM and low-altitude radar sensing systems. Other augmentation equipment can include target identification, GCI tracking, variable radar-image size, augmented infra-red radiation, and smoke system. Positive electronic identification is provided by the L-band IFF beacon.
Variants:
BQM-34E. For US Navy: total of 116 built. First operational flight 1 June 1972; in service from 1973 with Squadron VC-8 in Puerto Rico. Deliveries completed.

BQM-34E Firebee II, with ventral fuel pod

The illustrations on these two pages emphasise the slim form of the Firebee II (left and above) compared with the subsonic Firebee I (right)

BQM-34F. For US Air Force: total of 99 built. Slightly larger and heavier than BQM-34E, due to different augmentation and scoring systems and addition of recovery parachute assembly for mid-air retrieval system. Deliveries completed.
BQM-34T. Designation of US Navy Firebee IIs delivered since mid-1974 and fitted with Motorola integrated track and control system (ITCS). Total of 54 built; deliveries completed.

USAF BGM-34C MULTI-MISSION RPV
DANGER R.F. ENERGY

TELEDYNE RYAN MODELS 234 and 259 (BGM-34)

Tactical RPV; first flight 1973 (?)
Data apply to BGM-34C
Power plant: One Teledyne CAE J69-T-41A turbojet engine (1,920 lb; 871 kg st)
Wing span: 14 ft 6 in (4.42 m)
Length overall: 28 ft 6.2 in (8.69 m)
Body diameter: 3 ft 1.2 in (0.94 m)
Max launching weight: 5,000 lb (2,268 kg)
Performance: No details known
Launch and recovery: Air-launched from DC-130 Hercules director aircraft. Assumed to have parachute recovery system similar to Firebee I.
Guidance and control: Sperry Univac radio command guidance system and onboard TV transmitter.
Special equipment: Some BGM-34Bs fitted with extended, modified nose containing low light level TV camera and a Ford stabilised laser designator/receiver system weighing 250 lb (113 kg). This nose section is rotatable up or down to keep target in view during an attack. Equipment, which is similar to that contained in Pave Knife pods carried by F-4D Phantoms for use with laser-guided "smart" bombs, enables BGM-34B to act in pathfinder role, locating and locking on to a target and signalling its position to other RPVs, carrying weapons, in the force. One other BGM-34B was fitted with a Hughes high-resolution FLIR (forward-looking infra-red) nose sensor instead of the TV installation.
Variants:
BGM-34A (Model 234). Initial version (two built), with 1,700 lb (771 kg) st J69-T-29 engine. Trials included the release of single Shrike anti-radiation missiles, Maverick TV-guided missiles, and HOBOS homing bombs, all with good results, against mockups of Soviet Fansong SA-2 ground radars and other targets. Nose-mounted TV camera and Sperry Univac radio command guidance system.
BGM-34B (Model 234A). Development of BGM-34A, with more powerful engine, modified tail unit, enlarged control surfaces and greater operational capability. Eight built, for evaluation by USAF in weapon-carrying and other roles. Max launching weight 3,230 lb (1,465 kg). Trials included single and multiple passes against a variety of targets, launching a number of live and inert weapons which included SPASMs (self-propelled air-to-surface missiles) and Hughes AGM-65 Maverick TV-guided missiles. Has also demonstrated overland reconnaissance capability while being directed from a US warship standing offshore.
BGM-34C (Model 259). Multi-mission RPV, with modular nose sections for reconnaissance, electronic warfare and air-to-ground strike. Five ordered for USAF evaluation, by conversion of Teledyne Ryan Model 147SC (YAQM-34U) RPVs, with structure strengthened for ground launch. All delivered by December 1976. Capable of carrying double the weapons payload of the BGM-34B, including four Mavericks. This version combines, in interchangeable modular noses, the capabilities of the AQM-34V electronic warfare, AQM-34M reconnaissance and BGM-34B TV-guided Maverick strike RPVs. Lear Siegler avionics, with computer-based quick-turnaround automatic guidance electronics (AGE), are those originally installed in the YAQM-34U version of the Model 147 (which see). Flight test evaluation being carried out in 1976-77. Control aircraft is a specially-modified DC-130H Hercules, able to control up to eight RPVs simultaneously.

Teledyne Ryan Model 259 (BGM-34C)

Remarks: Developed from the Model 124 Firebee I and Model 147/AQM-34, this RPV is intended primarily for tactical strike and other defence suppression roles, reflecting plans to evolve combat RPVs for a variety of missions which at present require manned aircraft.

Far left: *Shrike missile (starboard) and Mk IV bomb (port) under the wings of a BGM-34A.* Left: *Rollout of the first BGM-34C.* Above: *Pathfinder version of BGM-34B, with laser designator and low light level TV camera in Ford nose pack* (Henry Artof)

(USA) TELEDYNE RYAN MODEL 235 R-TERN (YQM-98A) (COMPASS COPE R)

High-altitude long-endurance strategic RPV; first flight 17 August 1974
Power plant: One Garrett-AiResearch ATF 3 (XF104-GA-100) turbofan engine (4,050 lb; 1,837 kg st)
Wing span: 81 ft 2½ in (24.75 m)
Length overall: approx 38 ft 4 in (11.68 m)
Height overall: 8 ft 0 in (2.44 m)
Wheel track: approx 7 ft 6 in (2.29 m)
Weight empty: 5,600 lb (2,540 kg)
Payload for 24 hr mission: 700 lb (317.5 kg)
Max T-O weight: 14,310 lb (6,490 kg)
Cruising speed at 50,000-70,000 ft (15,240-21,340 m): Mach 0.5 to 0.6
Max endurance: 30 hr
Launch and recovery: Conventional runway take-off and landing, using retractable tricycle landing gear. Fully-automatic landing capability, in conjunction with Singer-Kearfott Talar IV microwave landing system.
Guidance and control: Teledyne Ryan radio command guidance system. Integrated flight control system incorporates computer, radar altimeter and ATC transponder. Aerodynamic control by ailerons, spoilers, elevators and rudder.
Special equipment: Equipment compartments in nose and tail portions of fuselage. Main mission payload compartment in lower forward fuselage, with provision for additional payload in rear fuselage. Details of other systems and equipment not released officially, but assumed to be similar in character to those of Boeing YQM-94A (which see). Alternative payloads tested in 1975.
Remarks: One of two contenders in USAF's Compass Cope programme (see entry for Boeing YQM-94A). Two prototypes built for 1973-75 evaluation, during which, on 3/4 November 1974, one aircraft set an unofficial endurance record for RPVs when it remained airborne for well over 24 hr and reached an altitude of more than 55,000 ft (16,765 m).

YQM-98A Compass Cope R photographed during its first flight, with landing gear extended

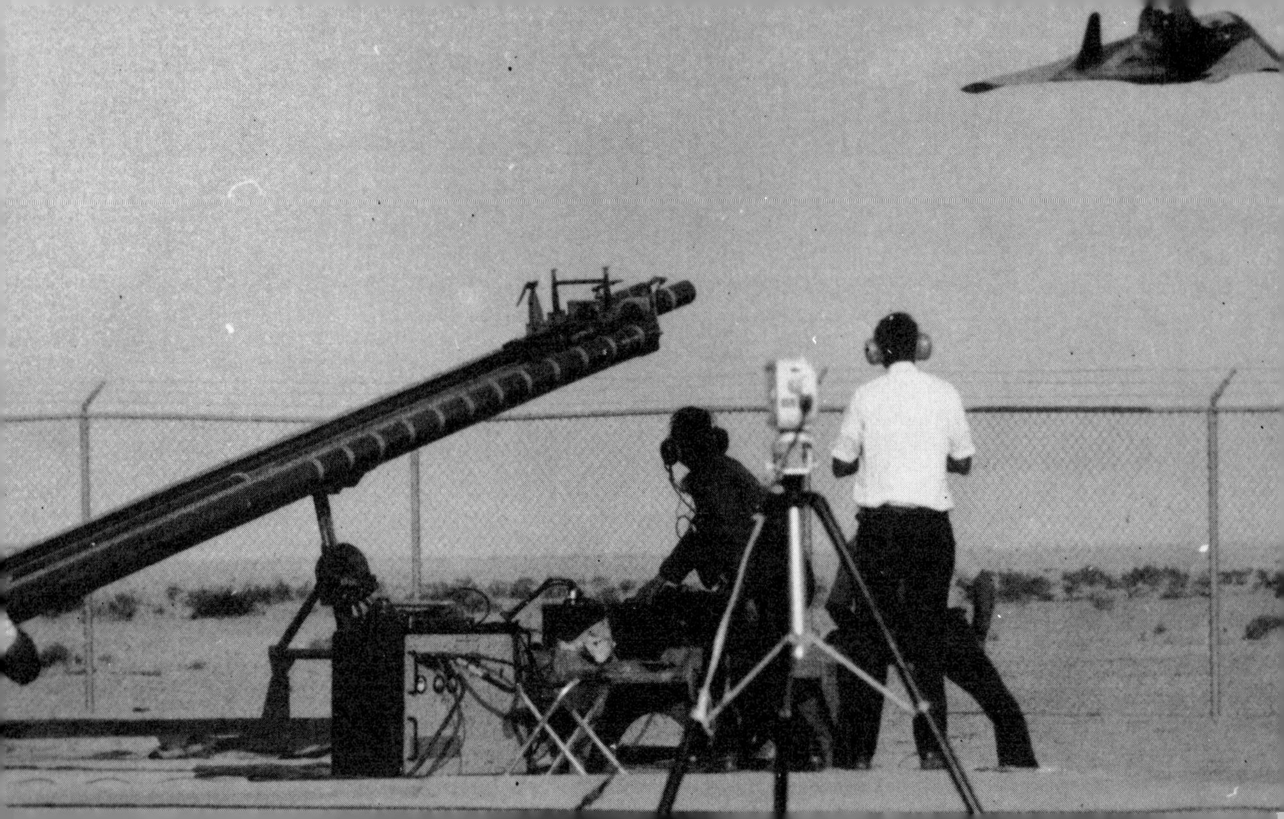

(USA) TELEDYNE RYAN MODEL 262

Experimental tactical mini-RPV; first flight 1976
Power plant: One McCulloch MC-101 flat-twin piston engine (25 hp)
Wing span: 10 ft 0 in (3.05 m)
Max T-O weight: approx 165 lb (75 kg)
Endurance: up to 8 hr
Launch and recovery: Shipboard launch from trainable compressed-air rail-type launcher; net recovery. May also be fitted with landing gear for conventional runway take-off and landing.
Guidance and control: Radio command guidance system. Hybrid autopilot, employing analogue techniques for the inner control loop and digital techniques for the outer control loop.
Special equipment: Modular payload packages, reported to include Honeywell electro-optical sensors, Naval Research Laboratory electronic warfare sets, and Raytheon radar.
Remarks: Developed originally for US Navy, for whom three evaluation models were built. Land-based tests completed successfully in 1976. First shipboard tests due in 1977. A scaled-down "feasibility version", with 7 ft 6 in (2.3 m) wing span, underwent testing in 1974. The delta wing is specially designed to contribute to a very low radar scope image and low visual, acoustic and infra-red signatures. Engine and fuel tank are mounted entirely within the aircraft, the former driving a five-bladed ducted propeller at the rear. Interest also reported to have been shown by US Air Force and Israel.

Teledyne Ryan Model 262 STAR mini-RPV being launched during tests at the US Navy's Naval Surface Weapons Center

GREENWALT

ARPV (Advanced Remotely Piloted Vehicle)
Intended successor to the Teledyne Ryan BGM-34C (which see), for service in the early 1980s. Designs, invited in 1975 from three competing companies, will incorporate a drone control and data retrieval system (CDRS) and data relay by Compass Cope-type RPVs. The CDRS, aimed at demonstrating simultaneous control of up to 20 RPVs in the face of intensive jamming, entered the prototype stage in 1974. Three-year study contracts were awarded in April 1975, resulting in competing designs from Boeing Aerospace, Northrop Corporation, and Rockwell International. The Northrop contender has a 15 ft (4.57 m) wing span, is 30 ft (9.14 m) long, and has a General Electric J85 turbojet engine. Rockwell's contender is of sweptwing, Vee-tailed configuration, and its power plant is installed in a dorsal pod aft of the wings.

TEDS (Tactical Expendable Drone System)
Original plans to issue RFPs (Requests for Proposals) for a TEDS vehicle to two competing contractors in October 1972 were deferred when proposed funding was deleted from the FY 1974 budget request. The TEDS programme was reintroduced in FY 1975, and the flight demonstration and validation phase began in early 1976. This is using, as recoverable demonstration vehicles, three converted Beechcraft VSTT prototypes (Model 1089E) and three Northrop Model NV-130s (MQM-74C targets, modified); the electronic warfare packages are designed by Sanders Associates and E-Systems respectively. Contractor proposals were submitted to AF Systems Command on 24 February 1975, and contracts were placed in May and June 1975. The Beechcraft and Northrop vehicles are intended as proof-of-concept demonstrators only, and are not necessarily contenders for a production TEDS. The latter will, of course, be non-recoverable, and is intended primarily for mobile ground launch, although air-launch capability may be included as an option, as in the Northrop NV-130 "Nutcracker". Payload will be in the order of 50-100 lb (23-45 kg), and primary electrical power source 1kW or less. TEDS is seen as having two basic functions: as a straightforward decoy, carrying active or passive ECM, or to accompany strike aircraft and provide them with ECM support. Luneberg lenses will provide the primary means of radar augmentation.

Artist's impression of the projected Rockwell International Advanced RPV

LCEHV (Low Cost Expendable Harassment Vehicle)
Mini-RPV programme to develop a small vehicle (possibly pre-programmed rather than remotely piloted) able to loiter over hostile air defences for up to 4 hr, either in an unarmed capacity to draw the defences' fire and provoke radar silence, or armed with homing devices and an explosive charge to destroy a hostile radar. About a dozen US aerospace companies were invited to complete paper studies in early 1975. System studies, by E-Systems' Melpar Division and Lockheed Missiles and Space Co, were submitted in mid-1975, and other contenders include Northrop and DSI. Flight testing of vehicles from these contractors began in Autumn 1976. Decision on future of programme expected in Autumn 1977. Meanwhile, field tests have been carried out with the E-Systems Axillary mini-RPV (which see), which has already demonstrated its radar homing capability. Another possible LCEHV contender is a 10 ft (3.05 m) long vehicle with a 14 ft (4.27 m) wing span, able to carry a 75 lb (34 kg) payload and two 2.75 in rockets. The US Army has a broadly similar programme for a small "kamikaze" RPV able to fly a warhead into a target.

USAF
Beechcraft
TEDS
KO-6E
MODEL 1089 E

Left: *Modified Model 1089E entered by Beech for the USAF's TEDS programme.* Above: *Model of the Rockwell ARPV.* Right: *Northrop VLCEHV harassment RPV, with its launch truck*

US NAVY XBQM-108A

Experimental RPV; first flight (tethered) 29 September 1976
Power plant: One Teledyne CAE XJ402-CA-400 turbojet engine (660 lb; 299 kg st), fitted with vane-type thrust vector control system
Main wing span: 7 ft 3 in (2.21 m)
Main wing area: 21.74 sq ft (2.02 m^2)
Foreplane span: 3 ft 7½ in (1.10 m)
Body diameter: 1 ft 2 in (0.36 m)
Length overall: approx 11 ft 9 in (3.58 m)
Fuel load: 94 lb (42.6 kg)
Max T-O weight: 560 lb (254 kg)
Max cruising speed: 400 knots (460 mph; 741 km/h)
Time to 5,000 ft (1,525 m): 8.8 sec
Endurance: approx 30 min

Remarks: Airframe modified from Northrop MQM-74A (which see), major changes involving the fitting of cropped-delta main wings, with elevons, fixed canard foreplane surfaces, a much larger fin and rudder, and a non-retractable tricycle landing gear from a Bede BD-5J. In this airframe are installed the engine (with a thrust vector control system added) and the midcourse guidance system of a McDonnell Douglas AGM-84A Harpoon air-to-surface missile. Emergency recovery parachute also fitted. Programme conducted by David W. Taylor Naval Ship Research and Development Center (NSRDC) and aimed at demonstrating VATOL (vertical attitude take-off and landing) capability. VATOL tethered test programme (without foreplanes) completed satisfactorily by early 1977; free-flight (non-tethered) testing due in late Spring or early Summer 1977.

XBQM-108A RPV photographed in tethered hovering flight. The aircraft's nose was, in fact, pointing vertically upward at the time this photograph was taken, with the umbilical cables leading to a tower

Wisp
Westland Wisp

WESTLAND WISP and WIDEYE

Mini-RPVs
Data apply to Wisp
Rotor diameter (each): 5 ft 0 in (1.52 m)
Body diameter: 2 ft 0 in (0.61 m)
Body depth: 1 ft 4 in (0.41 m)
Remarks: Westland Helicopters Ltd began work on RPHs (remotely-piloted helicopters) in 1968; experimental work began in 1972 using models for wind tunnel work and the measurement of radar, optical and infra-red signatures. At the same time simulated flights were made using a computer to check the stability and controllability of the RPH. Much of this groundwork had been completed by 1974, and was followed in mid-1975 by the start of a flight development programme, using a small flying testbed aircraft known as the Mote to prove the aerodynamic characteristics and to continue development work on the control system. This led to a Ministry of Defence development contract for the Wisp, a small RPH for the British Army. By Spring 1976 Westland had been selected as winner of the competition to develop a second RPH, known as Wideye, for battlefield reconnaissance for the British Army. The Wideye also is being developed under MoD contract.
Variants:
Wisp. A small number of Wisps are being built for trials purposes. The third of these was displayed publicly for the first time at the Farnborough air show in September 1976; flight testing was under way later in that year. The Wisp has a small, flattened-sphere-shaped body, made of glassfibre, which contains the power plant and mission payload. A fixed, four-legged landing gear is fitted to the underside. The power plant drives a pair of two-blade co-axial counter-rotating rotors; directional control is by differential changing of the collective pitch of the rotors. The Wisp can be carried in a standard Land-Rover vehicle, and a variety of interchangeable payloads can be installed.
Wideye. Wideye, at the design definition stage in late 1976/early 1977, is a larger and more advanced RPH, capable of carrying out a wide variety of operational roles. It too has a co-axial rotor system, a similar remote control system, and low radar, noise and infra-red signatures. The Electro-Optical Systems Division of Marconi-Elliott Avionic Systems Ltd is electronics systems contractor for Wideye.

Far left: *Third of the small batch of Wisps built for trials.* Left: *Model showing general configuration of the larger Wideye*

361

(USSR) YAKOVLEV Yak-25RD (?)

NATO reporting name: *Mandrake*
Target drone; all data estimated
Power plant: Two Tumansky R-9 turbojet engines (each 5,730 lb; 2,600 kg st dry and 8,818 lb; 4,000 kg st with afterburning)
Wing span: 70 ft 6 in (21.50 m)
Length overall: 51 ft 4 in (15.665 m)
Wing area: 538.2 sq ft (50.00 m^2)
Remarks: The original *Mandrake* was a high-altitude reconnaissance aircraft developed from the Yak-25 fighter, with an existing fuselage married to a new straight wing of extended span. This made possible the high-altitude performance needed for "U-2 type" missions, with a minimum of new design effort. The fuselage was believed to be generally similar to that of the Yak-25R/Yak-26, with a single-seat cockpit and a new "solid" nose to house reconnaissance equipment. The zero-track main landing gear of the Yak-25 was retained, with single nosewheel and twin-wheel rear unit under the centre-fuselage, requiring the provision of housings for the outrigger wheels on the tips of the new wings.

Photographs of a generally similar aircraft, referred to as the Yak-25RD, have appeared in the Eastern European press; this is almost certainly the "Yakovlev RV" which set a number of payload-to-height records in July 1959 and a women's closed-circuit distance record in September 1967.

The aircraft in the accompanying photograph was shown as a target for an SA-2 (*Guideline*) surface-to-air missile in a Soviet film made in the late 1960s, and appears to be almost identical with the *Mandrake*. This suggests that some *Mandrakes* may have been adapted into high-flying pilotless targets when retired from active service in their original role. It is also possible that some may have been adapted to continue their original role in a remotely piloted form.

This still of a target aircraft based on the Yakovlev Yak-25 fighter was taken from a Soviet film showing surface-to-air missile units in action

(USA) DSI RPMB

Developmental Sciences Inc has carried out flight tests with the RPMB (Remotely Piloted Mini-Blimp), a helium-filled miniature airship carrying an electronics package, propulsion system and ciné camera. The RPMB is 16 ft (4.88 m) long and has a diameter of 4ft 1 in (1.26 m). DSI is developing the project in collaboration with law enforcement officials and the FAA, and planned to have a 40 ft (12.9 m) long, 12 ft (3.66 m) diameter system operating by late 1977.

(Japan) JEAA FLYING SUBMARINE MODEL 503

This small, remotely controlled pilotless research airship was built by a group of members of the Japan Experimental Aircraft Association, led by Mr Daisaku Okamoto and Mr Kikuo Koizumi. It is powered by two 4.99 cc engines (each 0.63 hp), and is 25 ft 7 in (7.80 m) long. An unusual feature is the 4 ft 1 in (1.25 m) span, 9.47 sq ft (0.88 m^2) area wing fitted beneath the gondola to assist in vertical control and provide additional lift. A man-carrying version is planned.

DSI Remotely Piloted Mini-Blimp (RPMB) experimental lighter-than-air drone

(USA) US NAVY HASPA

HASPA (High Altitude Superpressure Powered Aerostat) is a programme for a remotely piloted airship for high altitude (68,000 ft; 20,725 m) ocean surveillance and communications relay, sponsored by Electronic Systems Command of the USN and managed by the Naval Surface Weapons Center. Martin Marietta is a prime contractor. HASPA will be helium-filled, with an electric propulsion system, will be 333 ft (101.50 m) long, and will have a diameter of 67 ft (20.42 m). An unpowered prototype, built by Sheldahl Inc, was due to fly in 1976.

N74170
EXPERIMENTAL

DRONE CONVERSIONS OF RETIRED FIGHTERS

Left: *FSI QF-86E, a drone conversion of the Canadair Sabre 5 produced for USAF evaluation by Flight Systems Inc. Like the other four aircraft illustrated in this final section of the Pocket Book, the QF-86E represents an economical way of producing high-performance targets or combat RPVs by modification of retired squadron aircraft*

Above: *One of many F-80A/C Shooting Star fighters converted into QF-80A and QF-80F drones, for use as missile targets and to fly into radioactive clouds to collect fallout samples*

52957
U.S. AIR FORCE

Above: *QF-4B target, converted from an F-4B Phantom II fighter by the Naval Air Development Center (*Howard Levy*)*

Left: *QF-104 target drone conversion of a YF-104A Starfighter*

*One of 29 F-86H Sabres converted to QF-86H drone configuration by the US Naval Weapons Center (*Henry Artof*)*

INDEX

T

U

V

W

X

Y